D1741507

THE NEW THEORY OF CONSCIOUSNESS

Delanin Fadahunsy

authorHOUSE®

AuthorHouse™
1663 Liberty Drive
Bloomington, IN 47403
www.authorhouse.com
Phone: 1-800-839-8640

First published by AuthorHouse   11/04/2011

ISBN: 978-1-4567-8767-7 (sc)
ISBN: 978-1-4567-8768-4 (ebk)

Printed in the United States of America

www.delaninfadahunsysconsciousness.co.uk

This theory has, throughout this compilation, employed the 'oddity hypothesis' paradigm, which infers that any attribute of an organism is deductible, inferable, observable or traceable from phylogeny or otherwise it is unnatural or insane; on the basis of which its claim of self verification is hinged, since the paradigm is like a piece of detective work in which it is impossible to plant evidence, unless our phylogeny is to be re-written.

Dedication

This book is dedicated to Dr. Peter Fenwick, the fiery luminary and daring lone voice in the wilderness of our ignorance on sleepwalking (somnambulism).

This 'New Theory of Consciousness' is a breakthrough in observing the natural world and in deductions that has decoded our consciousness in all of its states, and in so doing, has identified that in the natural state of consciousness called sleepwalking, it is impossible to commit violent aggression.

Sleepwalking murders appear to be rising over the years.

According to the BBC, (quoted in Wikipedia), there are now 68 known cases of Homicidal Somnambulism. The first ever reported case was in 1846.

'When they arrive in Court, the defense against murder charge has been on the order of 'I was sleepwalking and therefore ladies and gentlemen of the jury, I was not myself at the time I murdered (her/him) . . . . and so deserve acquittal.
    Culled from "Can sleep walking be a murder defense
    Lawrence Martin. M.D., FACP, FCCP
    Associate Clinical Professor of Medicine
    Western Reserve University of Medicine, Cleveland
    (Sleep Journal: Online June 2011)

'The fundamental legal argument used was based on the evidence of Dr. Peter Fenwick, who has elaborated his belief elsewhere (43) that somnambulism expresses a "disease of the mind". We do not agree with this opinion and would not wish to see patients with an episode of somnambulism with violence incarcerated in an institution for the criminally insane.'

From *R. BROUGHTON et al.: Sleep. 17(3):253-264©* 1994 American Sleep Disorders Association and Sleep Research Society *Medico Legal Issue. Somnambulism: A case report.263*

This breakthrough theory makes undeniable the fact that the act of sleepwalking itself is, much like sneezing or epilepsy, an naturally occurring non-insane automatism, its only function being to prevent sleeping mammals from falling through the *consciousness-continuum* into coma, hence any purported use of that state for aggression, is not possible and merely an excuse to conceal premeditated murder.

This new discovery proves that Dr. Peter Fenwick was right all along.

ODE TO CONSCIOUSNESS

Life from naught, a bitter fight for consciousness
A titanic struggle to wake from the abysmal sleep of death
And not slip back to whence it 'surrected'*
But in spite of best efforts at first
'Sleep' back it did, only to again re-commence the struggle
To lift up the Olympus within the veil of the darkness of death
To resurrect, again and yet again unto the daylight of the living

The hard victory over sleep—abysmal* it must defend
At any cost by staying alert
As even a wink, eternity may result
Trees ever at attention the wisest, refusing to sit or blink
Reptiles for motion must rest
And wisely sleeps while fully awake
Mammals only with abandon, sleep at own peril

Summary: This self-verifying new theory by observation and deduction, identifies that consciousness originated from nothing(zero; at the origin of living cells) and progressed in a gradual continuum through plant (vegetative) to reptiles to mammals to humans, illuminating how organisms attained those different levels, including the 'Permanent Access Memory (PAM) Hypothesis' in gaining self-consciousness. This theory discusses that mammals can only sleep at the level of reptilian consciousness (the sleep-wake consciousness) as the mammalian and human consciousness levels are sharp, focused cognitive states at which sleep is impossible. Rapid Eye Movement Behavioural Disorder (RBD) is the first safety net preventing mammals, at sleep, from falling through the *consciousness-continuum** into coma (vegetate consciousness) or death, and sleepwalking is the last vital safety net, if and when RBD fails; because even though a necessity for sleep, the reptilian consciousness is unsafe for mammals, due to the danger of sleeping through the *Consciousness-Continuum* to coma. It has now revealed, quite unexpectedly, the fact that in the state of consciousness referred to as sleepwalking, any act of aggression is impossible. Key word: *Consciousness-Continuum*.

This theory, a decade in the making is still dynamic and will undoubtedly remain so, for a while longer, as further attempts, are made to fine-tune it, as we try to unravel any remaining details of perhaps the most important aspect of our existence, our consciousness. If it has succeeded in, at least, laying a solid foundation, on which this can happen, then it would have achieved its objective. Updates, additions, discussions or contributions are available and welcome on *www.delaninfadahunsysconsciousness.com*

This book is made concise by design, to enable all level of readership keep track of the trend of events leading to our attainment of self-consciousness as well as the other issues central to its theme, without getting lost in details. A robust and detailed edition is expected to follow.

Contents

Title i

Dedication iii

Excerpts .... iv-v

Summary vii

Content ix

Preface xiii

The New Theory Of Consciousness 1-4

Preamble... 5-7

The Origin of Consciousness 8-9

The origin of sleep Non-Discernible Concentration

    Consciousness. 10-12

Sporadic Discernible Concentration

    Consciousness 13-17

The origin of sleepwalking 18-20

Permanent Discernible Concentration

    The Origin of Self Consciousness..................... 21-46

The Three Tiers Hypothesis of Consciousness 47-57

The Hypothesis 58-63

Sleep state of consciousness 64-70

Dreams state of consciousness 71-73

Sleepwalking state of consciousness 74-86

Legal Perspective (mens rea) 87-95

Consciousness Compulsion 96-98

Brief word on Julian Jayne 99-103
Conclusion as Positional Synopsis...........................104-138
Bibliography 139-146
Glossary..147-148

Preface

CONCIOUSNESS

Consciousness is the insignia of life, without which no life exists. Awake or sound asleep, consciousness is the absolute determinant of the presence, absence or the ebbing away of life, the ultimate distinguishing factor between the dead and the living. From Aristotle to Descartes and considered by many philosophers through the ages, consciousness or what really constitutes consciousness has remained as enigmatic as it was elusive. Even with advances in science and medical fields, the problem of clearly defining the meaning of consciousness or its origins had defied logic. For as long humans have been self-aware, they have also sought answers to the mysteries of the How, the When and the Why of consciousness with questions such as:

What is and what constitutes consciousness?

What or where is the origin of consciousness?

What is and what constitutes self-consciousness?

What really, if any, is the difference between consciousness and self-consciousness?

What or where exactly is the mind?

Are animals as self-conscious as humans?

What actually are dreams? Senseless images or potent messages?

What is sleep and what is the origin of sleep?

Why do we sleep and what is REM (Rapid Eye Movement) actually all about in sleep?

What is the sleep cycle and why do we need to have it, or what does it connote?

What is sleepwalking and why do people sleepwalk?

Can crimes actually be committed during sleepwalking?

How do we know if and when people hide under the cloak of sleepwalking to commit a crime?

Do animals sleepwalk?

What is coma?

The concepts of Consciousness are now decoded by Delanin Fadahunsy (DF) and answers are discussed herein to all of these age-old questions and more.

The New Theory Of Consciousness

Evolution of consciousness: The hypothesis of surrection

Life surrected from death, which we can refer to as zero consciousness, and evolved from there in a gradually increasing continuum, the Consciousness-Continuum (CC), through different stages. The stages of the continuum include broadly single–celled plants, simple animals, complex animals and sea creatures, amphibians, reptilian, mammalian to the highest level that is human consciousness. The order of the consciousness continuum:

Surrection→Sporadic Attention→Permanent Attention →Non-Discernible Concentration→Sporadic Discernible Concentration→Permanent Discernible Concentration

Synopsis of the evolutionary history of organisms

From substances in the "primordial soup" (biochemical synonym for puddles from which the earliest living organisms formed) to simple cellular entities (able to use primitive form of photosynthesis or chemosynthesis to survive) to eukaryotic (complexity within the cell) cells, to multicellular sea creatures to amphibians to reptiles to mammals to humans.

Table 1. Theory of Consciousness

TYPES OF EXISTENCE	DEGREE OF CONSCIOUSNESS	CREATURE TYPES	TYPES OF CONSCIOUSNESS	CLASS	INTENSITY FOR MEMORY ACCESS	TYPES OF ACCESS TO MEMORY	FUNCTION	SLEEP TYPE	SLEEP SIGNAL	SLEEP JUNCTIONS	EQUATION OF LIFE
Death / Surrection	O Ascending(MIN)	-	Non-consious ESS	-	- (MIN)	- (MIN)	-	DEATH / SURRECTION COMA--	NORMAL ALARM DANGER ZONE	(MIN)	(-)
Formation of Life	From > O to > O ascending	RNA/DNA	Barely consious	-	- (MIN)	- (MIN)	-	COMA--			
First Lives	From ascending > O to Sporadic Attention ascending	Up to First surviving cells	Sporadic Attention consciousness	-	Ultra Ultra Low	None	-	COMA +			
First complex lives	From Sporadic Attention ascending to Permanent Attention ascending	Up to Algae, Plants	Permanent Attention consciousness (PA)	3rd Class Lower	Ultra Low	None	Non-cognitive	COMA			
First mobile lives	Permanent Attention to 'Non-Discernible Concentration' (NDC)	Up to Reptiles	Non-Discernible Concentration consciousness (NDC)	3rd Class Upper	Low	None Ritualistic	Non-cognitive Ritualistic	NREM+SWWALK			
Cognitive	NDC ascending to Sporadic Discernible Concentration (SDC) ascending	Up to Mammals	Sporadic Discernible Concentration consciousness (SDC)	2nd Class	Medium	PVMA HPVMA	Cognitive	REM			DF EQUILIBRIUM OF CONCIOUSNESS
Cognitive / self-reflective	Permanent Discernible Concentration	Humans	Permanent Discernible Concentration consciousness (PDC)	1st Class	High	Sporadic/ Permanent	Cognitive / self-reflective	FULLY ALERT			(MAX)(M)(M) Human consciousness (+)

SPORADIC

NREM

DFS DREAMS — SLEEP CYCLE — DFC

LEGEND
PVMA - Partial Virtual Memory Access
HPVMA - High Partial Virtual Memory Access
DRS - Delusional Virtual Memory Syndrome
DFD - Delusional Virtual Memory Disorder
CNC - DF/VNR
0 - O Consciousness or Death
+ and ++ Deeper and Deepest state Coma

*Figure 1.*DF Equilibrium of Consciousness (DF EQcons)

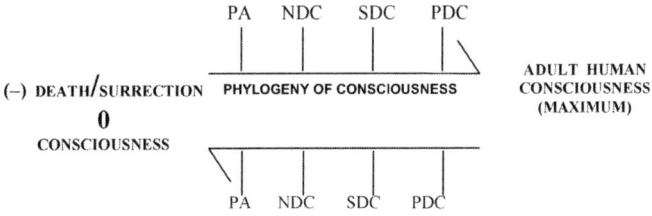

LEGEND

PA   - Permanent Attention Consciousness

NDC - Non-Discernible Concentration Consciousness

SDC - Sporadic Discernible Concentration Consciousness

PDC - Permanent Discernible Concentration Consciousness

Life is consciousness in a delicate equilibrium, the direction of which is influenced by anyone or the resultant of all the internal and external stimuli effective on an organism, with phylogeny onwards in the forward aspect and backwards in the reverse.

Preamble

Delanin Fadahunsy's *Philosophy of the surrection of Life*

Life at the very beginning of evolution started in fits and starts, a 'punctuated equilibrium' and not necessarily in a smooth sequence. An imaginary ageless person at that beginning would have seen a barren wasteland, with no hope of any life ever being possible. Even when the seasons began and rains started, the person was convinced that life from the silence of death was impracticable, with no movement, no sound except for the rains and oceans.

After a long time covering millions of years, the ageless individual passed near a puddle, a' primordial soup', and out of the corner of the eye thought there was a movement in there and looked again in the puddle, but saw nothing. Only fleetingly visible, the titanic struggle for life's surrection from death which had been going on at the subterranean level for these millions of years was beginning to yield results. It was to take another long period, many millions of years before the individual was certain that something pulsated from that 'murky puddle', and not long afterwards, the pulsations came more often. After the first surrection living entities struggled to survive, propagated and experienced death. The formation of life was truly underway. Against all odds, life had beaten death to attain consciousness. The hard won struggle for consciousness from the unfathomable depth of the sleep of death was robustly defended. Non-alertness threatened life and was the anti-thesis of life and, sleep, the height of non-attentiveness.

Table 2.
Synopsis

Type of Existence	Degree of Consciousness	Life form
Death/Surrection	0	
Formation of Life	From 0 to >0 Ascending	RNA/DNA
First Lives	From >0 Ascending to Sporadic Attention Ascending	First Cells
First Complex Lives	From Sporadic Attention to Permanent Attention Ascending.	Algae, Plants
First Mobile Lives	Permanent Attention to Non-Discernible Concentration 'NDC'	Up to Reptiles
Cognitive Life	NDC Ascending to Sporadic Discernible Concentration Ascending	Up to Mammals
Cognitive/ Self-Reflective Life	Permanent Discernible Concentration	Humans

The basic timeline, with very approximate dates

5.8 billion years of simple cells (prokaryotes)
3 billion years of photosynthesis
2 billon years of complex cells (eukaryotes)
1 billion years of multicellular life
600 million years of simple animals
500 million years of arthropods (ancestors of insects, arachinids and crustaceans)
550 million years of complex animals
500 million years of fish and proto-amphibians
475 million years of land plants
400 million years of insects and seeds
360 million years of amphibians
300 million years of reptiles
200 million years of mammals
150 million years of birds
130 million years of flowers
2.5 million years since the appearance of genus Homo
(Source: Wikipedia Online Encyclopedia; March 2011)

Origin of Consciousness—
Permanent Attention
Consciousness—Algae, Plants

Life at the beginning was fiercely competitive for very scarce resources in the intense duel for survival, which meant that coming to existence was only just the beginning of the struggle for any of the species, and surviving long enough to produce offspring was just as equally crucial. For life to be preserved in the midst of such fierce contests it was not enough for an organism only to fight to survive, it had to also remain alert or attentive to its surrounding, for apart from anything else, it could perish in an instant, by having a vital resource or resources taken away by a fiercer opponent.

Therefore, attentiveness to one's surroundings became a selection pressure, and moreover, it soon became obvious that if remaining attentive was very important, sporadic attention may not be good enough. The longer an organism stayed attentive, without wavering, the better its chances of survival. Since the struggle for survival was day and night, organisms that stayed attentive night and day survived. Organisms that could not cope with this level of attentiveness did not survive. Therefore permanent attention of an organism to its surrounding became so vital that it was now engrained into the very fabric of life itself and became its insignia.

Sporadic Attention and Permanent Attention Consciousnesses

Permanent attentiveness to one's environment is the same thing as being permanently conscious of it; therefore, if sporadic attention was considered the nucleus, then permanent attention is the atom of what we term consciousness, and henceforth can be considered its origin. Therefore, we may now say, up until this time in evolution, two types of consciousnesses had emerged, the first of which is Sporadic Attention Consciousness, which died out very early in the timeline; and secondly, Permanent Attention Consciousness, exhibited by algae, plants and other similar organisms to date and which forms the basic core of consciousness for all other levels of consciousnesses above this. Therefore, in higher consciousnesses, Permanent Attention Consciousness forms the basic layer upon which other successive levels of consciousnesses are formed. In other words, no conscious or living organism exists without Permanent Attention Consciousness, or the possibility thereof, in one form or another. It is our most primordial inheritance, our instincts and everything underlies this principle.

The Coniferous Coast Redwood is the tallest tree species
on earth (Source: Wikipedia Online).Trees are a prime
example of Permanent Attention Consciousness

'Non-Discernible Concentration' (NDC) Consciousness— The Reptilian Consciousness and the origin of sleep.

As the explosion of life continued, competition for scarce resources grew more intense and became so fierce that dispersion became inevitable to allow advantage to be taken of greener pastures elsewhere. With dispersion came increased risks, as organisms had to be even more wary, suspicious of, and more vigilant in new environments. Moreover dispersion meant the need for motion which also meant increase in energy level required for living, and as life grew more active, due to the need to move further afield, the greater became the need to find a way to incorporate a modicum of rest in the new lifestyle; as opposed to the formerly sedentary existence such as algae types and photosynthesis-based lifestyles.

The more active life became, the greater the need to meet the new challenges it brought, and therefore a never-resting, always-alert, year-in-year-out, way of life was no longer realistic, if the organism was not to die of exhaustion or become unable to survive long enough to reproduce. This pressure eventually resulted in the growing of an appendage that would allow the need for the highly desired increased vigilance that movement caused, and at the same time allow some measure of rest, while also meeting other needs of the new challenges. That appendage appeared in certain organisms and with time, grew and became the reptilian brain. With this, the organism can now have it both ways: the possession of a heightened level of attentiveness, as well as a fairly more relaxed state of existence that allowed rest, enhancing rather than compromising the security of the creature in the intermediate world that lies between sleep and wakefulness. This is the origin and beginning of the phenomenon called sleep.

This is the world afforded by the reptilian brain, and a creature here literally lives its dreams, because its sleeping state is also its wakefulness state. However, the new heightened level of the Permanent Attention now made possible by the new reptilian brain is high enough to be likened to some form of low-level concentration, which even though it is higher than permanent attention, is still too low to be discerned as 'concentration', and will herein be termed, 'Non-Discernible Concentration' 'NDC'. Up until now in the evolution of consciousness, we had 'sporadic attention' which then became permanent attention which by virtue of the reptilian brain became the highest level of attentiveness attainable in the form of NDC, which borders on the next stage up, which is 'discernible concentration'. Thus, the reptilian brain has raised the consciousness level to 'Non-Discernible Concentration Consciousness' 'NDC'.

Coast garter snake *Thamnophis elegans terrestris*

An example of Non Discernible Concentration (NDC) consciousness. NDC operates on Non-Access Memory (NAM) mode, procedurally like an automaton, since there is no cognitive memory (Intuitive/Reasoning).

'Sporadic Discernible Concentration' consciousness (SDC)—The Mammalian

As speciation continued, and life became ever more active, the sphere of existence widened more considerably, necessitating much further travelling afield or migration, which meant entertaining the risk of the unknown that constitutes new environments. Therefore if NDC was so far adequate in fairly familiar environments, where life was lived rather casually, procedurally and routinely, it was a different situation when migrating much further into totally strange and unknown environments. In the first instance, just being in such environments alone meant being wary and fearful, even if there were no obvious dangers, and so organisms whose movements have so far being fairly smooth and almost mechanical in their somewhat familiar environments suddenly became erratic. Every noise or stirring caused by even the wind demanded a quick response in the execution of the 'fight or flight', a situation that may not have warranted such actions in their previous fairly familiar environments. Furthermore, at those instances of intense fears of the unknown, the attentiveness of organisms became more intensified than if such was happening in the former environments. Thus, inadvertently the hitherto Non-Discernible Concentration (NDC) gradually adapted to Discernible Concentration. However, such intensification or Discernible Concentration only occurred at the instances of those panicky fears, but the acute concentration remained in place until the danger was gone, and so Sporadic Discernible Concentration (SDC) consciousness came into existence. Furthermore, the extensive nature of migration meant organisms needed prior knowledge of places they have been before and the dangers associated with such spaces. They also needed to be able to flee from places they have visited before and which they intuitively remembered for its dangers. This interplay of different types of fears is the stimuli required for the further development of the reptilian brain, since the brain, as empirical evidence revealed herein later showed is a stimuli-reactive organ. The more it was prompted by these different stimuli, based as they were on organisms' primordial instincts of survival, the more it grew to accommodate such needs. It is also only logical to assume and thus understand that the brain structure as at this time in evolution still had the potential for growth or what might be called 'the existent option of growth' on the evolutionary menu. In time the brain, due largely to experience in the environment and survival, grew like a 'waiting-to-rise' freshly kneaded dough to meet the various causative challenges. Such challenges met include the great navigational abilities of the birds or the fishes migrating to and returning from far and near places where they had been compelled for myriads of reasons such as, among others, safe reproduction and rearing of their offspring or periodically escaping certain unfavourable conditions. There was also the inclusion of the very vital, hitherto non-existent, part with information storage potential, memory, in the new brain. Thus organisms no longer rely on procedures or rituals for existence, but could actually call up information, albeit sporadically, when the situation demands, to resolve their problems, thus endowing them with a brain far superior to the former reptilian.

Organisms now had two level of consciousness by this time in evolution, the first of which is NDC and the second of which is SDC, supported respectively by the old reptilian brain, whose functions was reduced to basic issues, and the new mammalian brain on which they became reliant. They both work hand in hand, for example when there is no danger apparent, organisms relax the guard to operate on the lower level consciousness of the reptilian brain, and when danger lurks or other situations such as excitement or other stimuli call for it, they operate on the SDC. This meant that when an organism is at its highest alert, it has switched from NDC to SDC, at which point, all potentials available are ready for use, including the best, which is memory. However, SDC is only able to give them a level of concentration sporadically, which is only to allow memory usage, as the need arises. At all other times, memory access is automatically shut down. Therefore, whenever the concentration level is not up to discernible level, memory access is rendered impossible; and to re-access memory, a full level of 'discernible concentration' is required, as such memory access is always sporadic.

With such capable brains, organisms felt more and more secured, able to exploit the new faculties to their advantage in detecting threats or prey, spatial cognition, understanding how to secure themselves, as well as defend their territories more efficiently; gradually resulting in less reliance on the reptilian brain, which was now used only for basic functions. The reliance on a more focused brain grew greater; which in turn meant that organisms gradually lost the original brain, and with it, its sleepy, dreamy, in-between, nature; to the sharper, more focused and intensely cognitive nature, afforded by 'discernible concentration' of the new mammalian brain.

Bottle nose dolphin; an example of SDC consciousness
Organisms with SDC operate on Random Access
Memory (RAM)

The ability to store and use information whenever needed meant organisms could, or example, file data on what surroundings look like, so they knew what to expect, and if no changes or cause for alarm, reduce guarded behavior, with no further need to sleep while keeping vigil (fully awake) to have a totally peaceful rest, rather than live in a state of never really sleeping, a half-way reptilian world between sleep and wakefulness.

The origin of sleepwalking (somnambulism)

However, this situation presented an unexpected problem, which is that sleep is the anti-thesis of consciousness, and therefore a dangerous phenomenon for organisms because sleep, the deepest form of sleep, from which organisms will lose all forms of consciousness, is death or *sleep abysmal*. Therefore originally the reptilian adaptation to sleep, when it became necessary, as discussed above, appeared to be the safest answer, sleeping while awake, thus removing any risk of sleeping to death. Consequently, even though mammals now have a sharper and less android—like type of existence, they still needed to sleep, which was impossible in their new state, unless they went back to the reptilian state of consciousness, since consciousness is a continuum. So, switching from this state to mammalian state of consciousness presented a problem, which took organisms time to adjust to in a gestation period that meant gradually shedding their surreal reptilian consciousness for the sharper mammalian consciousness, but in which the body chemistry had to adjust to sleeping without being awake. Initially, in order to eradicate the problem of a sleep-wake state, which is actually a state of sleepwalking, organisms on falling asleep had to be immobilized. This was achieved by the paralysis of muscles movements in sleep, through the suppression of the neurotransmitters responsible for neuron movements. This is the origin of the condition known as REM atonia.

However, as already made apparent, sleep portends one of the greatest dangers to organisms, especially sleep at which all consciousness, but the very basic, is lost, as distinct from the sleep of a sleep—wake safe mode. This is because in this unsafe mode, the Consciousness-Continuum (CC) hitherto tied to a sleep-wake mode, is now untied, able to freewheel; whereas, in sleep, as consciousness drops, there is only one direction the CC could freewheel, as dictated by the DF EQcons, which is towards coma and death. Therefore, when an organism's consciousness has dropped too much, when its sleep becomes too deep, such that its consciousness is getting to the point where it could fall through the CC into coma, it then is compelled to get up and re-enact the reptilian mode of consciousness. This is the origin of the phenomenon known as sleepwalking or somnambulism. The act of walking around in the old mundane robot-like existence of that state of consciousness, serves to firstly instantly halt the organism's descent through the CC into coma, and secondly, to exercise the organism to raise the level of consciousness up before it is then returned back to sleep; and since consciousness is a continuum, no sooner than the organism returns to sleep than it is headed yet again in the wrong direction of the DFEQcons, and the process repeats, culminating in what is referred to as the sleep cycle, but which is really a consciousness-balancing cycle, herein referred to as the DF cycle.

'Permanent Discernible Concentration' Consciousness— The Human

As the unstoppable march of evolution continues, an even further expansion of the mammalian brain was compelled, in order to allow permanent access to memory. However, this was due to a near catastrophe, in the evolutionary history of the Hominin, when individuals adopted the posture habitual bipedalism.

Bipedalism briefly defined, is the act of using the hind limbs for locomotion by an organism that has four limbs, while using the forelimbs only occasionally for the same or other purposes or permanently only for other purposes. The Early Hominin, EH, favored the latter style, which then exposed individuals to unprecedented levels of predation. While all of the 'brother' ape lines wisely used that posture only occasionally, the EH refused to, which is in the theory referred to as 'postural habitual biped fervour' (PBHF), and because of this was easy prey. The predation level on the species got so bad they were close to the edge of extinction. As a result, the EH was forced to start acquiring a longer span of concentration, wary of the ever presence nature of death, in the stealthy, virtually never ending attacks of the formidable cat class.

ORIGIN OF SELF-CONSCIOUSNESS

Self-consciousness, more than likely, came about in the last stages of the evolution of the human brain from the mammalian brain; most probably out of a near-perilous event in the evolutionary history of the ape-line of human ancestry, the EH, when bipedalism, a scare stunt became adopted as a habitual posture, instead of occasional.

Bipedalism

Bipedalism, according to a number of arguments which this theory supports, conferred no significant advantages on the EH and its appearance at first was without any obvious particular adaptive benefit. Being an important link in this theory, the advantages and disadvantages of biped are briefly revisited.

Pros

- A fully upright position, by mammals or reptiles is an intimidating posture that strikes paralyzing fear into prey, making it easier to subdue, and in the same vein could scare away threats. A bear standing up to full height, in an attack, gives such an intense dose of fear that its work at subduing any threat is that much reduced. Similarly, an upright cobra strikes fear into prey, especially when the creature advances towards a target with that menacing posture. Therefore, the fully upright position by mammals, as conferred by bipedalism, is almost undoubtedly, a 'scare stunt' that most probably, originated instinctively, and was retained only because of the effectiveness as such.

- The 'free' forelimbs (relieved of their primary locomotive duties) were supposedly used for food gathering, female provisioning (Lovejoy Co. Science 1981) and in making tools by the extant apes.
- It was a sexual selection tool.
- It conferred, on the EH a greater field of vision into its environment, to spot prey or a threat more quickly.

Cons

- The always fully upright position of habitual bipedalism, offers preys and predators similar ability to easily spot the EH from afar and when this is put alongside the better night vision of say the cats, then, a habitual erect position, just so to see far, can be more costly than gainful. Hence, it might have brought more harm than good, to the EH, since it must have compromised the family position time and time again, exposing them to unprecedented level of predation attacks.
- Furthermore, taking perhaps the most important of the senses, sight, away from where the action is, into mid-air, on a permanent basis, in a bitter war of survival, could be considered a disadvantage for the EH, against those ground bound, well camouflaged enemies.

- Postural habitual bipedalism would appear to be contrary to the logical uses of almost all of the natural mammalian senses of sight, smell, taste, touch and hearing, which were all adapted for ground use. Taking these away from their natural position, on a habitual basis could definitely erode their effectiveness, say for example, the sense of smell, with resultant consequences.

- While an upright position may be an effective 'scare stunt', there is no evidence to suggest that it is anything more that that, since the venom of an upright cobra is hardly any more potent in that position than if it was delivered from the ground; nor does it make the charge of a bear any less effective.

- Arguments that it conferred speed for attack or defense are flawed in that the high centre of gravity makes maneuverability poor, in a cunning rat race for survival; which in turn makes catching prey or escaping threat that more difficult. As a matter of fact, it probably conferred lethal disadvantages, opening the gangling EH, more and more to relentless predation attacks; most especially from the agile cat class (c-class), whose four legged posture on a streamlined body combined with a lower centre of gravity allows for better body handling, superb maneuverability, and stability, especially at speed. Timed running speed of the fastest biped, the ostrich has been estimated at about 65km/h

(40 mph), while that of the cheetah can exceed 100km/h (62mph), (Sharp N.C. (1994).

- The forelimbs represent perhaps higher mammal's (apes) most formidable tool of escape; in fact much more so, especially in a wooded habitat, than any consideration of the factor of speed, since they were naturally adapted for 'arboreal gymnastics escape' (age); which is far more important when trying to get away from a more agile opponent such as the c-class. A careful consideration of this point makes any argument for speed against such creatures look ridiculous; for to compare bi—to quadraped is, in the instance, like comparing a two to a four wheel drive in a race for survival on a rugged terrain.

On the contrary, an ape could easily escape a big cat, by its arboreal gymnastics, which are almost totally forelimbs dependent. The forelimbs of the habitual biped, used mainly for food gathering, looses its biceps and other muscles' power for lack of the exercise such as derived from habitual locomotion. Constantly flexing the palm of the forelimbs, which is closed when moving and open at all other times also helps gripping ability, which effectiveness is also lost in habitual biped. As a consequence the forelimbs loose their effectiveness in the type of leap to grip or move from branch-to-branch in myriads of direction, vertical, horizontal, or combination fashion, to escape a threat. Even the best of the c-class, agile as it is, would be a laugh for an ape, as it (the cat) clumsily climbs (climbing ability is limited mostly to the trunk or main branches of) a tree, while the ape can make its get-away even if it caught a dried branch that snapped, because the other forelimb was already catching a better one; as it, almost effortlessly, swings away from danger. In fact, the loss of this capability alone, would have been suicidal and possibly the greatest singular failure brought on by the adoption of habitual bipedalism, and it must have put the EH at a great disadvantage and very drastically increased predation pressure on similar species of apes.

- Arguments about stone tools being developed for defensive or attack purposes, with the forelimbs appear unrealistic, as these tools are most probably ineffective against big, cunning predators moving at considerable speeds. Besides, there is no known record of use of such tools at this timeline in evolution. (Winifred Henke and Thorroff Hardt1997)

- It is such a compromising position in an environment, where crouching offered better survival chances, that it would only have left the EH practically defenseless and vulnerable; in fact so much so, that it is logical to assume in such circumstances the probability that the c-class actually preferred and specifically hunted down the EH species, as energy-saving easier caught prey, enabling it to quite considerably decimate their population; to such an extent that it actually could have influenced hominin evolution(Winifred Henke and Thorroff

Hardt 1997). In this sense, it is quite conceivable that the c-class possibly nearly once hunted our ancestral line of apes to near extinction.

- On the whole, it seems fairly clear, that indeed bipedalism most probably had no obvious adaptive purposes attributable to its sudden appearance, other than being an effective scare stunt, which came about intuitively; and therefore ought to have been used only very rarely for when that purpose arose, as observed for the cobras, while keeping a low profile at all other times.

In the 'Handbook of Paleoanthropology' Winnifred Henke and Thorroff Hardt (1997)' posited that:

'Bipedalism predates use and manufacturing of tools and probably did not evolve in the open . . . as was once assumed and it predated robust Hominin dental complex, hunting and social structure changes, which could be related to provisioning females with meat food. The hunting hypothesis (Brain 1981) can be applied as a selection force only for late Pliocene Hominins. EH were rather the hunted than the hunters and this also means that predation pressure on EH in late Miocene and most of the Pliocene could have been quite strong and therefore could have influenced hominin evolution. The main predators of our ancestors were probably the leopard (PantheraPardus), Sabre tooth felids and possibly hyenids. Only very rarely in extant apes are such behaviours as throwing objects manifest or bipedal display; thus such factors are very unlikely selection pressures to posture habitual bipedalism.

The cost of running for the type of bipedality used by EH was probably higher than the cost of walking;

therefore possible selection for Homo such as biped was rather an unlikely pressure for emergence of biped. Biochemical arguments for supporting hypothesis relating to morphological, biomechanical,and or physiological responses to the physical environmental influences and inextricably related to open habitat, cannot be accepted for the origin of bipedalism in EH, but rather be applicable to the new adaptation that probably appeared only in the genus Homo (e.g. the loss of elaborate thermoregulation by eccrine glands)'

And A, Haviland, HalandE.N.Prins, Dana Walrath and Barry McBrd in 'Evolution &Podostomy;the human challenge'(1998) surmised that:

'Fossil, flaura and fauna found with the Arditpithecus and the possible human ancestors from the Miocene are typical of a moist closed, wooded habitat. However, the presence of bipedalism in the fossil record without a savannah environment does not indicate that bipedalism was not adaptive to these conditions; it merely indicates that bipedalism appeared without any particular adaptive benefits at first. We now know not only that bipedalism preceded the evolution of larger brain by several million years, but we can now consider the possibility that bipedalism may have pre-adapted human ancestors for brain expansion.'

Sexual selection may be responsible for bipedalism

In the context of the fact that bipedalism probably conferred no significant advantages, and might have been a curse rather than a blessing, what then could possibly have been responsible for its appearance and upkeep? This theory posits that while it might have originally started as merely a scare stunt by the EH that yielded results, in which case it became occasional; in addition to that, sexual selection

might have been one of the motives for the increase usage of that posture. Apart from being a psychologically superior position over both foe and love, it does expose the manhood to the female directly at the front, giving it much more prominence as a whole part. This issue does not necessarily have to be considered from the view point of phallic display (Tanner 1981: 165) to be effective on the female

The fastest land animal, the cheetah can reach 120 km/hr (75 mph)

Hominin, because, in the case of humans for example, a woman's breasts displayed provocatively, with or without erect nipples, is more effective a sexual tool than when not so displayed or partly concealed. In addition to this, an erect gait is far more masculine than the crawling position, another fact that would not have been lost on the female Hominin. All of these reasons point in the direction of sexual selection as more than likely being the major pressure responsible for bipedalism.

Postural habitual biped fervor (pohabifer) hypothesis

However, while sexual selection may very well have been one of the pressures for occasional bipedalism, it is very unlikely to have been responsible for habitual bipedalism; as if it were so, there could possibly be no plausible argument for why it could not be used occasionally, if for that purpose only, during mating periods or seasons; especially when juxtaposed with the survival dangers it portended in the disadvantages aforementioned. Furthermore, if the EH used it as a sexual selection tool, it is reasonable to assume that other extant apes used it for the same purposes and the fact that they did not go on to adopt it as a habitual posture, will suggest that using it occasionally, would have been quite adequate for sexual selection purposes.

Therefore, there had to be another reason for the EH's seeming determination to make that posture habitual, under those circumstances. There is one logical direction to look, which is that, a peculiar trait of the EH might have been the reason for its unwillingness to make it an occasional posture. The position of this theory is that the EH is, when compared to other primates, specially predisposed to the motivational senses of its mammalian brain.

Therefore, once the EH stopped crawling, it somehow suited this predisposition. The superiority that the posture potentially confers not only fitted with its persona, but also 'locked' onto that persona. Therefore, it is this trait that would be held solely responsible for the appearance of posture habitual bipedalism and is

An ostrich, one of the fastest of living bipeds

herein referred to as "posture habitual biped fervor", since the other ape lines, most likely used it for all other reasons successfully, and did not adopt it permanently. Positions such as that it begun as a fashion that just caught on (Dawkins e.g. 2004) may be untenable in that apart from appearing too weak an argument, (without a kind of 'stubborn streak' peculiar trait, being the primary reason; the EH would have gone back crawling like brother apes when the need arose—in the likelihood of danger), it is suggestive of an attempt to project human behaviors on extant apes.

The 'Permanent Access Memory' (PAM) hypothesis of the attainment of self-consciousness.

In a self-created predicament, a reversal of which is impossible, habitual bipedalism having now been 'locked' onto its persona, the EH must have become like a cornered animal from the onslaught of the cat class now directed specifically at the specie, as they were easier to spot from a distance and easier to catch. The same motivational trait or stubborn streak that gave rise to habitual bipedalism, in the first place, would also not allow the EH to give up that posture, regardless of the cost. It seems once our ancestor got off the ground, he was never going back crawling again.

It would appear that the line of pre-humans were peculiarly naturally motivated and strong willed enough to refuse to go back to crawling again, no matter the consequences. However, the cat class was also relentless in its attacks, the sudden many—fold increase of which must have caught the thoughtless EH, by total surprise. As such the EH, more than any other line of apes, started living quite unexpectedly in mortal fear of the cats and was forced to become much more attentive than any other ape line to its environment. This extraordinary circumstance for the EH and the fear created by an ever present nature of death, more specifically among its specie meant that such acute attention became a common occurrence. But as earlier deposed to by this theory, acute attention to one's surrounding is the same thing as discernible concentration to the environment and as such the EH's span of discernible concentration gradually grew longer as the cats' ceaseless onslaught continued. This meant that the EH more than any other line of apes, hiding away in terror at each attack, watching helplessly as the cat class's gory slaughter and consumption of its family members continued, became ever more acutely attentive in concentration and since these are the signals or prompts, the pre-requisite for the growth of the 'ready to rise' young brain, its brain grew. The increase in discernible concentration span is the origin of self consciousness. This is because being able to increase concentration span beyond the sporadic meant the barrier to memory access which hitherto had remained sporadic, is being broken down and memory is now becoming more accessible. The combination of memory and concentration constitutes self consciousness, but only when it is beyond the sporadic or what may be referred to as emergency or random access. Memory access is the same thing as presence of mind. Therefore, in emergency access, such as the only access possible in SDC, memory is only present to resolve an issue and no longer available, even in the slightest fashion, after the problem is solved. Therefore, presence of mind is so fleeting it is non-existent, since it is only for a purpose, which is to solve a problem and not for anything else. However, while this is true for all organisms up until now in evolution, the EH unwittingly was moving the barrier further and further. While sporadic concentration was adequate enough for all other organisms to notice a threat and execute the 'fight or flee', with an approximately even chance of success or failure for both predator and prey, such was no longer the case for the EH, because adopting a permanent biped posture (originally unnatural, being learnt and needing getting used to, with 'teething' problems as in any new habit) had drastically reduced its chances of success of escape as prey, in favour of the then de-facto ruler of planet earth, the cat class. As the spans of its discernible concentration increased, so did its mental capabilities. This is because this theory holds that intelligence is much less a function of the size of an organism's brain than it is the ability to utilize memory. As such the ability to utilize memory is related to the intensity and span of concentration attainable by such organism. Only concentration at enough intensity (Discernible Concentration) is able to open the door to the memory of an organism. This theory further holds, as previously disposed that access to memory equates presence of mind. Thus when an organism concentrates, even if sporadically, it enjoys a presence of mind for that particular period; reason why animals are able to play games as complex as those by bottlenose dolphins, such as in rhyme with human counterparts. Since only Discernible Concentration is able to access memory, animals are only able to utilize memory at the instance of the right stimuli, e.g. when their interest is ignited, in danger or other such needs, and straight after the task, their level of concentration naturally drops back to Non-Discernible Concentration level, and they are thus rendered incapable of accessing or utilizing memory, until another such situation that can ignite concentration to Discernible level occurs. Therefore, the ability of the EH to start utilizing memory, with gradually but steadily increasing presence of mind (Discernible Concentration) and the resulting growth of the brain in response to the usage meant increasing mental capabilities. Thus as the cat class's onslaught continued, hiding and watching the cat slaughter and consume loved ones turned to attitude, such as howling, running around wildly to distract the predator's attention from continuing with the attack or consuming the prey in peace and other such antics. All these were, as to be expected, logically ineffective against a hungry cat, but with each failure came more trials, and gradually these wild uncoordinated antics became more tactical, as the EH mental capabilities kept increasing, with each trial and error, since the problem persisted. Afterwards, as a matter of the same course of events, the tactics advanced to the development of rudimentary tools for deployment against the cats. The interplay of these factors continued up until the time that the EH's span of discernible concentration (presence of mind) and consequent mental capabilities enabled it to subdue the cat class. As earlier discussed in the positions, mentioned above, of Winnifred Henke and Thorroff Hardt (1997), this theory believes that the period for these events lie approximately between the late Miocene and most of the Pliocene, estimated at between 5.332-2.588Million years ago (Mya) (Source: Wikipedia Encyclopedia Online, March 2011) when it was that the predation pressure on the EH was so strong, it might have led to extinction. Therefore an estimated guess of this theory of the period when humans ancestral line of apes gradually extended their discernible consciousness span from the typical mammalian SDC, which was when self consciousness began to develop, will also lie within that period. It is therefore safe to assume that by the end of the Pliocene, approximately 2.588Mya, human ancestors had gained a fair degree of self consciousness, but were not fully self conscious, otherwise the chances of surviving the cat class as habitual bipeds, was next to nil. This is because by being bipeds, they constituted themselves into the number one food source for the then top of the food chain, the cat class, and because they could not match the cats in physical and ferocious prowess, only brain power could have saved them from almost certain extinction. Therefore, by this timeline, the EH already possessed something none of the other ape lines, nor indeed any other organism had ever had; a much extended span of discernible concentration which meant access to memory was better and superior usage of that memory was becoming apparent, as the mind became more present. It is possible that it took another considerable period, probably a couple of million years more before full self consciousness was gained, when discernible concentration then became a totally permanent feature. Paleoathropological evidence shows that not long after the period that the EH was supposed to have managed to subdue the cat class, early humans have advanced in their tool making abilities. This is because the *Homo habilis,* whose fossil was discovered in Duvai, Tanzania in 1960 (Leaky et al 1960), and severally named the 'Handy man' or the 'Duvai toolman' (Leaky et al 1964), famous for his tool making prowess, lived in the period, around 1.75Mya (Source: The Archaeology Online Journal: archaeology. com June 2011). By this timeline, it is only logical to assume that the discernible concentration or consciousness span of our ancestors had increased considerably and so was their brain power. In comparison with other ape lines, they were already definitely self-conscious, enabling them to achieve such feats, in furtherance of their predecessors (EH)'s rudimentary abilities. From then on full self-consciousness was just a matter of time and only held back because the full large picture, or PDC, that makes up memory or mind, needed to be developed. This represents the entire database of an organism, which includes everything the organism knows about itself and the world. Therefore, by the time the Home sapiens came into the picture, 500000Mya (Source: The Archaeology Online Journal: archaeology.com June 2011). there was hardly any doubt that PDC was already fully developed and we were fully self-conscious. And when the brain, in keeping pace with the dynamic situation, had grown all it can to accommodate (reach an equilibrium with) the biochemistry of having concentration on a permanent, rather than sporadic basis of any duration; the modern human brain had fully developed and full self consciousness was born. Initially it may not have been too difficult for the EH to contain the cat class, mainly because it already has a database of knowledge (memory) of its foe on which it engaged its increasing span of presence of mind with the attendant level of intelligence attainable to subdue the cat class. However, gaining full self consciousness took a bit longer because, for it to be gained, such as we know it today, there was the need for the EH to develop the database of memory, which forms the PDC, beyond its immediate environment and problem. Therefore for most of the period between the estimated 2.588Mya and up until around the suggested 500000Mya, full self consciousness was still being developed. One of the hypotheses of this theory, holds that 'From ontogeny to adulthood, the homo sapiens, the highest consciousness, recapitulated the phylogeny of consciousness' Therefore, the fact that self-consciousness did not come all at once, but was gained only gradually can be seen today as a human child develops, increasing in self-consciousness only gradually, which is not fully attainable until young adulthood. Another evidence of the gradual nature of the development of self-consciousness is evidenced in the corroborations provided by Paleoanthropology, as regards this assumption.

In the timeline preceding bipedalism, the existing apes had smaller brains, but around the timeline that the EH's predicament began, brain size took a leap forward in growth, as shown below:

Timeline	Human Ancestors	Brain size
6-7Mya	Sahelanthropus tchadensis	350cc
3-3. 9Mya	Australopithecus afarensis	375-500cc
3-2Mya	Australopithecus africanus	420-500cc
2-1.5Mya	Australopithecus robustus	530cc
1.75Mya	Homo habilis	590-710cc
1.8Mya - 300,000ya	Homo erectus	750-1255cc
500,000ya	Homo sapiens (archaic)	1200cc
195,000ya	Homo sapiens (modem)	1350cc

(Source: The Archaeology Online Journal: archaeology.com June 2011)

Once the inertia of strictly sporadic access to memory was broken and that difficult barrier was crossed, full development to the human brain was truly under way and it was only a question of time, from then on, for its full development. This is because from then on, the more the young, waiting-to-grow brain was prompted or used, by the ever extending discernible concentration or presence of mind, the more it grew. As revealed in Table 5 above, between the period of the first listed ancestor, the *Sahelanthropus tchandesis* and the *Australopithecus afarensis*, covering about

3.1 million years, the brain grew only about 25cc, but from then on it took a leap to 530cc, almost doubling, by the time the EH, in the *Australopithecus robustus* timeline, was supposed to be out of trouble and had effectively contained the cat class, a period of only approximately 1.9 million years. In further support of the "Rising Dough' hypothesis of the growth of the human brain below explained, the brain took off from there, 530cc, and had an amazing almost three fold increase in about only 1.3milliion years to *Homo sapiens (modem)*, 1350cc. This is because as earlier explained, the mind or discernible concentration was becoming ever more present, and the corresponding memory or PDC usage, which represents the prompts or stimuli needed for the relatively young brain to grow, increased many folds, enabling the 'rising dough' of a brain to complete its rise quickly. Furthermore, from the period of the Homo erectus and the Homo sapiens (archaic), 1200-1255cc, the brain grew a paltry 100cc to 1355cc (Homo sapiens-modem), in a period of about a million years. The only logical conclusion that may be drawn from these is that it has by now grown (risen) all it practically could.

Empirical evidences abound demonstrating beyond reasonable doubt the presumption that the brain is capable of growing with a discernible level of concentration. The most recent of which is the Sara Lazar experiment in which she demonstrated that the practice of mindful meditation can make measurable changes in the brain regions associated with memory, sense of self and empathy and stress. (Source:

Harvard Gazette. A online Journal of the Harvard University, June 23, 2011)

It is important to understand that while these experiments are often truncated as demonstrations that the brain can grow due to such prompts, this theory believes that they are in fact nothing more that a demonstration that the brain once grew. This assumption is only logical, because it is generally assumed that further significant growth of the human brain is highly improbable, and yet there is hardly any argument about the fact that it grew from the original mammalian brain, in a relatively short period of time from between 3.9Mya to 1.5Mya. The only conclusion that may be drawn from these two facts is that espoused by this theory in the "Rising Dough hypothesis of the growth of the human brain", which is that the brain was once, in its younger years much like a knead wheat dough waiting for the yeasty condition (right prompts or stimuli) to grow, and once, like a fully risen dough, it has grown to size, is really incapable of any further significant growth, like it did in the past.

If the EH now had Permanent Discernible Concentration (PDC); it is in addition to the mammalian Sporadic Discernible Concentration (SDC), which it already possessed, creating a superb capability that can be likened to a 'picture-within-a-picture' television, where permanent discernible concentration is the larger picture and sporadic discernible concentration, the in-picture, whose focus shifts. The nature of this capability is such that even long after the EH, came through to this ingenious level of consciousness, and long after its immense capacities were put to use in defeating to start ruling the cat class, as well as all of living nature, the modern human was unaware of it, because it was gained so gradually. This same scenario can be seen as a child turns from mammalian to a human consciousness. When the individual gains self-consciousness sometimes in childhood; neither the child nor the parents are aware than something so fundamentally great, as an animal consciousness becoming human consciousness has occurred, and to all concerned, its just one other ordinary day. However, while the EH on attainment of self-consciousness could be imagined to put its incredible potentials to use immediately due to an already existing database of information in its mammalian brain memory; the human child will have to rely on SDC until it has any PDC, at all, in place for self consciousness. This explains the reason for childhood behaviours which means they always have to rely more on their mammalian consciousness as PDC gradually develops, thus acquiring self-consciousness only gradually, as they acquire the larger picture to compare with in the workings of the mind.

Humans (Homo sapiens) male and female (Source: Wikipedia Online 2011) The only PDC Consciousness known, are exponentially more intelligent due, not to a larger brain size, but, to a permanent presence of mind.

The Rising Dough (RD) Hypothesis of the development of the human brain

Earlier in the evolutionary timeline at which the EH's predicament occurred, this theory believes that the young brain was still highly impressionable waiting, ready and eager for the right condition to fire off and grow. If the firing of the neurons of our present fully developed brain is comparable hypothetically to a machine gun, then that of our earlier mammalian brain was comparable to a 'cluster bomb' with each firing resulting in many more other

Below a reconstruction of *Australopithecus afarensis*, a human ancestor that had developed bipedalism, but which lacked the large brain of modern humans

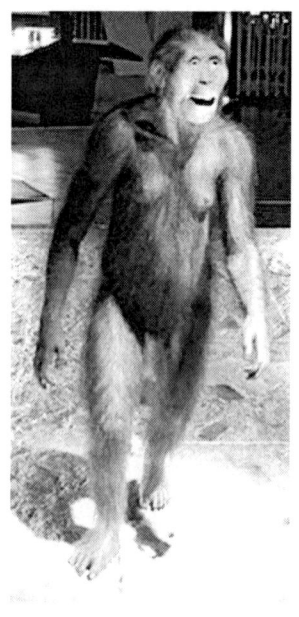

multi-lateral firings; enabling its full formation within a relatively shorter period, as above explained. The brain itself was then comparable to a freshly knead wheat bread dough, waiting for the yeasty condition to rise. The fact that there is conclusive evidence that our brain did develop from an earlier mammalian type brain supports this hypothesis and also serves as incontrovertible evidence that the option of further expansion existed then on the evolutionary menu, and the fact that any further significant development of the now fully developed human brain is highly improbable is further attestation to the RD hypothesis of the then 'freshly knead dough' having risen all it could now. The right condition for that growth was the atypical convergence of five factors which were (i) the existence of a peculiar stubborn trait in the EH, a specie of primates (ii) the manifestation of that streak as *postural habitual biped fervor*, (iii) the resultant (as a consequence of the sudden manifestation) unexpected and unprecedented relentless predation onslaught by a cat class discovering a 'new' specie of easy preys, the EH, thereby effectively pitching our ancestors with the then rulers of planet earth, the cat class, in a bloody, gruesome, winner takes all, duel for survival; but which unknown to both contestants is essentially a duel for the ownership of the blue planet (iv) the increased usage brought on memory by the resultant gradual increment in discernible concentration span of the EH and (iv) the possibility of the option of further brain development on the evolutionary tree. This theory further postulates that but for the presence of the latter, our ancestors could have been made extinct by the cat class and a wild earth would have been inhabited by all other organisms present today only, ruled by the cat class, in the likes of the lion king, but no humans. An earth without humans is not only imaginable, feral, clean, free of artificial synthesis, greed and pollution; it very nearly came to be.

The 'Control of Fire' (COF) Hypothesis

If self-consciousness originated with and because of the stated reasons around the 2.588Mya watershed, it is necessary to similarly define the point at which it could be deemed incontrovertible that humans could be said to have possessed full self–consciousness, the latter being a gradual process.

This theory defines the full self consciousness watershed as between 500,000 and (definitely not later than) 125,000 years ago, using COF as marker. It is generally believed that COF was a crucial moment in terms of its cultural significance in the evolution of humans, due to the possibilities it heralded such as food cooking and the provision of warmth that undoubtedly made possible or at least aided geographic spread into cold climes, among others. Much more than that however, this theory believes that COF established, beyond any reasonable doubt, the fact of a permanent presence of mind, or a fully established Permanent Discernible Concentration (PDC) in the timeline of human evolution. Fire is perhaps the most attention–commanding element known. A group of apes beside the white-sand and blue-ocean pacific shoreline, for example, will sooner notice a camp fire around than pay any attention whatsoever to the awe–inspiring grandeur of the latter, and any organism that has been able to go as far as control fire for use can be assumed, under even the most rigorous assessments, to have an established or permanent presence of mind, in order to achieve that feat, in the light of what this theory has propounded. It takes PDC to concentrate on fire, understand its properties enough to make use of it repeatedly and be cautious enough to try to avoid its dangers.

'Incontrovertible evidence of the widespread COF dates approximately 125,000 years ago and later. Evidence for COF by Homo erectus beginning 400,000 years ago has wide scholarly support, while claims regarding earlier evidence are mostly dismissed as inconclusive or sketchy. Claims for the earliest definitive evidence of COF by a member of the Homo range from 0.2 to 1.7 million years ago' (Source: Wikipedia Online Encyclopedia article on 'Control of Fire').

Evidences of COF by humans abound variously in Africa, Asia and Europe, around the period 500,000–125,000 years ago as suggested by this theory, a few examples of which, among others, are:

1. The Cave of Hearths in South Africa with burned deposits dated from between 200,000 to 700,000 B.P., as do various other sites such as Montagu Cave (58,000 – 200,000 B.P.) and at the Klasies River Mouth (120,000 – 130,000 B.P). At Chesowanja, in East Africa, archaeologists found red clay sherds dated to be 1.42 Mya. Reheating on these sherds show that the clay must have been heated to 400 °C (752 °F) to harden.

2. The Qesem Cave, 12km east of Tel Aviv, with evidence of the regular use of fire from before 392,000 to around 200,000 B.P. at the end of Lower Pleistocene. The large quantities of burnt bone and moderately heated soil lumps suggest butchering and prey-defleshing took place near fireplaces.

3. Zhoukoudian cave in China, with evidence of fire as early as 230,000 to 460,000 BP. Fire in Zhoukoudian is suggested by the presence of burned bones, burned chipped-stone artifacts, charcoal, ash, and hearths alongside *H. erectus* fossils.

(Source: Wikipedia Online Encyclopedia article on 'Control of Fire').

It is safe to conclude therefore, using COF as cut-off point, that from between the 500,000 to 125,000 years ago, humans were, beyond reasonable doubts, already fully self conscious.

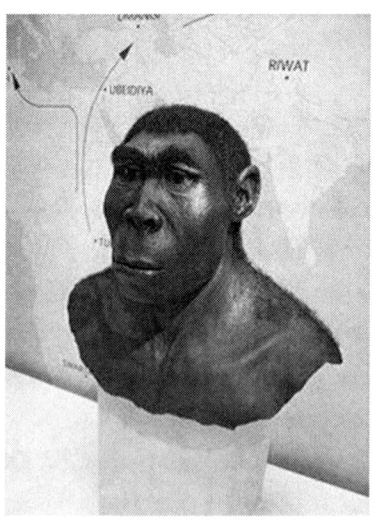

A reconstruction of *Homo erectus,* the earliest human species that is known to have controlled fire. *(Source: Wikipedia online encyclopedia; article on 'Control of Fire')*

The Three Tiers Hypothesis of Consciousness

While the other apes (and mammals) can also concentrate or brood, they can only hold concentration at the right level sporadically and for a span, which is whenever the mammal is faced with demanding situations e.g. when excited as observed in dogs or bottle nosed dolphins playing games or when faced with danger. And since concentration is the password for access to memory (as earlier posited); while the other mammals with only sporadic concentration, can only access memory sporadically (when situation demands); the new human brain now has permanent access to memory. Furthermore, the depth of concentration means, memories can be better utilized, by the new brain meaning that while the old brain is at best capable of say, approximately, a (human) child's ability, though in varying degrees, depending on the species, the new brain equates to a human adult's capability. A human child (herein meaning before self-consciousness is gained) is only conscious, possessing only sporadic discernible concentration, and not self-conscious. However the self-conscious human, an adult has both SDC and PDC; where the former represents attention and the latter is the mind (permanent access to memory). Thus a situation is created whereby PDC is always there like a large screen of a television, while SDC acts like the 'in-picture' which means that, even though the latter can shift focus as situation demands, at no time is the full picture lost. This capability is self-consciousness and self-reflection is the usage of such capability when for example the focus of the 'in-picture' shifts to the self, which also happens to be the self in the larger picture and a scenario of reflection, of self-looking-in-on-self, is thus created as self (SDC), the in-picture, looks into self in (PDC), the larger picture.

Therefore, this theory posits that there are three distinguishable classes of consciousness, the first of which is the lowest category (Third class), and represents just Permanent Attention at the lower end. All living things have third class lower consciousness which is a common denominator. It is basic and represents our most primordial inheritance; the first line of defense. Next to that level is Third Class Upper consciousness, which is the highest level of attention, too low to be discerned as concentration, but is as close as it gets to some low-level form of concentration (NDC) and is herein referred to as 'Non-Discernible Concentration'. Next to which is Second class consciousness.

This is discernible concentration, which while on a much higher scale than third class upper consciousness or NDC, is not permanent, but sporadic, referred to as SDC (occurring only when situation demands) and its duration varies from species to species and what task is at hand,

THE NEW THEORY OF CONSCIOUSNESS

among other things. This, of course is in addition to NDC, and next to that is First class consciousness. It is made of permanent attention (non-discernible concentration), sporadic discernible concentration and discernible concentration which is not sporadic, but permanent, (PDC) and on a higher scale than that available in second class consciousness; of course alongside NDC. Therefore, in summary, a creature in:

Third Class in two categories, upper and Lower or Non-Access Memory (NAM) consciousness

Third Class Lower consciousness or Permanent attention Consciousness has:

- Permanent attention only
  e.g. Algae, Plants

Third Class Upper consciousness or Non-discernible concentration consciousness' has:

- NDC (Non-Discernible Concentration) ONLY
- Too low an intensity for memory access
  e.g. Reptiles

Third class consciousness is present permanently, albeit, in varying degrees, in all living things. It is the first line of consciousness; our primordial instincts.

Second Class or Random Access Memory (RAM) Consciousness

- NDC + SDC (Sporadic Discernible Concentration)
- Sporadic access to memory

- Memory utilization when accessed not as good as in first class consciousness due to the lower intensity quality of concentration and its sporadic nature. e.g. Mammals

First Class or Permanent Access Memory (PAM) Consciousness
- NDC + SDC + PDC (Permanent Discernible Concentration)
- Permanent access to memory
- Brilliant utilization of memory due to permanence of access and the intensity of concentration attained. Humans only

Furthermore, this theory posits that concentration, whenever it occurs, opens the door to the memory of that creature. Therefore, what happens is that when a creature operating with second class consciousness concentrates (which is sporadic; only when needed), the memory access door opens for that brief period; the creature uses the opportunity of the memory access to retrieve information, with the ability to access randomly and use, before the door is shut and further access is denied, which is when the concentration or task at hand stops. Hence, second class consciousness may be referred to as 'RAM'—Random Access Memory consciousness, only available for specific periods to perform a given task, after which further access is denied, for loss of concentration, until next task comes up and the process repeats.

If only random access is possible in the second class consciousness level, because concentration is not permanent, then at the first class level, permanent access results due to permanent concentration; and therefore if second class consciousness is 'RAM', then first class concentration will be 'PAM', 'Permanent Access Memory' consciousness. In the same vein, third class consciousness will be NAM 'Non-Access Memory' consciousness, because the low level of concentration 'NDC' cannot open the door to memory (even if it were present). Finally, because concentration (the password to memory) is permanent in the first class level, the door to memory is always open.

Memory is everything a creature knows of itself, present, present-continuous or past, stored biochemically. Therefore, a creature in first class mode, with the door to memory permanently in the open position is always looking at its memory. The memory storage is the mind, which means it is always examining itself. This means that the brain has the capability of first generally over-viewing all that is in there and also into any part of the storage that is of interest at any time and because what is in there is an integral part of that creature, the memory or mind is only serving as a mirror, reflecting back the creature's self.

The permanent and innate ability (made possible by the new brain power) to look in through a permanently open door at our own memory, even if we are not looking at anything in particular, creates the permanent impression of looking in on one's self. In other words, permanent concentration when attained lights up the entirety of the organisms' being and self-awareness is created. Animals are biochemically 'wired' such that everything connects to the memory, and so any of them with permanent access to the memory is like a switch turning on the light, with which a creature with that capability can see all that it is, as long as the light is on.

However, whenever memory is turned off, then the switch turns off, such as sleep time, until memory is on again. It's akin to a recording camera that is always on, whether focused or not, until turned off.

To better illustrate the point, an animal does not possess self-consciousness as it can only see the part of the body or situation that warranted the induction (the stimuli); whereas a human (self-conscious) creature already is seeing the entire memory (due to permanent access).Therefore, when a stimuli induces a response, it will only need to turn attention (SDC or in-picture) to the position of that stimuli, while still holding the whole full picture (PDC or the larger screen). Therefore self-consciousness is not unlike a 'picture-within-a-picture' Television capability while on the other hand self-unconsciousness is like a blank screen (as the larger picture), with the 'in-picture' showing and shifting as the attention shifts, the larger picture lighting up sporadically only when situation demands (concentration high enough) and shutting down as soon as the inducting task ends. Therefore, in self-consciousness, it is possible to watch the current happenings in the in-picture knowing that the larger picture holds on permanently, so long as consciousness is not off (as in sleep). The act of looking at the in-picture while the large picture is permanently holding, for comparison, is self-reflection, because whatever is in the in-picture thus can be related to the larger permanent picture. This is not so in self-unconsciousness, where because concentration is sporadic, the phylogeny never made possible a permanent larger picture and all there would be is the in-picture on a blank screen, with nothing to relate it to. It is this made possible relationship (dual-imagery), in self-consciousness that creates self-awareness. The attainment of Permanent Discernible Concentration by the EH created the permanent larger picture (which is seeing permanently into our memory i.e. what we are), which is now added to the original in-picture already attained by the mammalian brain (Sporadic Discernible Concentration) to create self-consciousness. Self-consciousness is being aware of oneself (larger picture) in relation to the outside world, or a part or whole of oneself (in-picture).

This is what constitutes self-consciousness. Because our memory records as we do anything, we could see ourselves do (present continuous) that thing in real life and time recording (in-picture). So, we could see ourselves eat, run, or do anything so long as discernible concentration remained permanent and the memory door is open; we can now record and watch the recording as it is happening or take ourselves back and forth of the recording, while still holding the second larger image, which is permanent. Whereas sporadic concentration leaves only the possibility of an open (only to utilize) and shut (immediately afterwards) memory access (in-picture) and no more; the strength (intensity) and concentration of that level of consciousness is never enough for it to keep the memory door permanently open to enable a larger permanent picture and create a comparison or self-reflection.

Thus humans basically, like other animals, have intermittent attention (sporadic concentration) spans, (the in-picture), but the ever-present nature (except in sleep, when all but Permanent Attention, are switched off) of the whole (larger) picture (memory) prevents us from losing the continuum each time our attention shifts; providing the much needed back up to enable a continuous and seamless integration of our perceptions and thoughts, which is what is referred to as 'presence of mind (memory) without which we would not be self-conscious. In other words, the Permanent Discernible Concentration enables the integration of memory (the past) with the present continuous on a permanent basis, thus allowing for a projection, even if rough, of the future; while other mammals, with SDC (or the present continuous only and only sporadic access to the past) have only the immediate life (present continuous) to contend with. John Locke (1690) in his Essay 'Concerning Human Understanding', defined consciousness as 'the perception of what passes in a man's own mind'. It was, literally, as close as anyone could ever get, until now.

Are there instances when the door to memory, the mind, also slams shut on a creature capable of self-consciousness as a result of a break in permanent discernible concentration, PDC? The answer to this self-verification question would logically have to be yes, since it is only discernible concentration that is the password to memory (the mind);

therefore, anything that leads to loss of PDC for a period (as in sleep), means memory access is denied for that period and permanently means memory is lost permanently, even though the creature still has PA to fall back on. Here is an example of such a period, which is a true-life experience of this theorist.

A night out resulted in binge-drinking with friends, eventually culminating in my blackening out.

It was at this stage that I had to walk back home, a distance of about a kilometer away from the bingeing spot in an urban environment, with the necessity to cross roads, which mercifully had very light or non-existent traffic at 3 to 4 a.m. It is pertinent to mention that I was a very light drinker, and this situation was the first time I had gone this far; a total blackout that I never experienced before. The fact was that it was not until being forced to get up from bed at about 7.30 a.m. for some wrenching fits that I first re-discovered myself, at which point, *discernible concentration*, the password to memory, just returned (alcohol having lost its hold) and I started imagining how on earth I got myself home. This self-experience was invaluable because it gave a self-realization that I had crossed three main classes of consciousness.

Diary of events
- Got to the party in PAM mode, first class consciousness or state of both permanent attention, NDC and PDC—Full Human Consciousness.
- This switched to RAM mode, second class or Mammalian consciousness, at the point in the

bingeing, when I could no longer concentrate permanently, which meant I had lost PDC. I operated then like animals do, with permanent attention but only sporadic concentration, at and when necessary. And like any animal, I could take care of myself, as in fight or flee, or even play games, albeit drunkenly with off and on concentration; as I could still access memory, albeit sporadically when something calls for concentration. A person, unlike other animals is aware of and adept at using PDC with SDC, while an animal is even unaware of the existence of PDC, let alone be adept at its use, and can only use it instinctively, sporadically.

- Switched to NAM mode, third class consciousness, at the point when I no longer had any concentration, and the memory door had thus slammed shut, with access permanently denied, for as long as the alcohol held me in that state. Mind you, I still had NDC, or PA, which is common to any organism with life in it. More bingeing could only result in the erosion of the faculties of NDC, up to a point, in my case, where all I had left before leaving the party was in the form of procedural memory (like a sleepwalker) to guide me home. I could no longer fight or flee in case of danger, or even take care of myself, like an organism adept at using NDC, as for example a reptile could.

Another proof of the fact that humans possess mammalian consciousness, SDC, is in insanity, where even though the PDC (the larger picture) is lost, the SDC (the in-picture) remains and such person is as only self-unconscious as an animal.

The hypothesis of presence of mind (POM) over brain size as a measure of intelligence

By now, the conclusion is almost unanimous within the science community as regards the fact that neither absolute brain weight nor the ratio of brain to body weight provides sensible criteria for comparing intelligence in different species. The Table below not only shows the brain and body weight of different species of organisms but also their brain to body weight ratio, a scrutiny of which will attest to this assumption.

This theory posits that presence of mind is the criterion that determines intelligence in species with brains. Presence of mind is here defined as discernible concentration without which brain access, as previously explained, is impossible. Therefore, it follows that the longer and better (in terms of level of intensity attainable), the level of discernible concentration, the better the use of the available resources in the brain, which is the memory or database of an organism. It is therefore not surprising that humans with Permanent Discernible Concentration are the most intelligent and other mammals with only Sporadic Discernible Concentration are far less intelligent. Furthermore, this hypothesis adequately explains the reason for the big difference between human and animal intelligence, even though no such difference could be found in a comparison of either their brain sizes or the ratio of their brain to body weight. The table below shows, for example that, were brain sizes to be used as a measure of intelligence, then the bottlenose dolphin, Killer whale, Pilot whale, Sperm whale and Fin whale will all be far superior in intelligence to man; and the Sperm whale over five times as intelligent as man. On the other hand, if brain to body weight ratio were to be used as the measure, then the mouse will almost be twice as intelligent as man. Yet there is hardly any doubt that man is exponentially more intelligent than any of these organisms. The only reason that can be logically adduced for this, is the one proposed by this hypothesis, which is that man's ever present state of mind, allows for an exponential ability of memory usage.

Species	Brain weight	Body weight	Brain to body weight ratio
	(gram)	(ton)	%
Man	1500	0.07	2.1
Bottlenose dolphin	1600	0.17	0.94
Dolphin	840	0.11	0.74
Asian Elephant	7500	5.0	0.15
Killer whale	5620	6.0	0.094
Cow	500	0.5	0.1
Pilot whale	2670	3.5	0.076
Sperm whale	7820	37.0	0.021

| Fin whale | 6930 | 90.0 | 0.008 |
| Mouse | 0.4 | 0.000,012 | 3.2 |

Source (The High North Alliance brochure: 'LIVING OFF THE SEA' "Mink whaling in the North East Atlantic' February 1994)

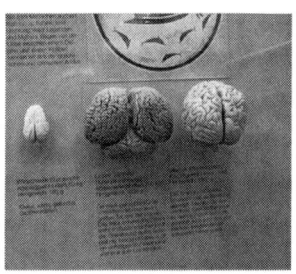

Scale model of bottlenose dolphin (*Tursiops truncatus*) brain (middle), compared with brains of wild pig (*Sus scrofa*) (left), and man (*Homo sapiens*) (right) (Source: Wikipedia Encyclopedia Online 2011)

The Age Hypothesis of Consciousness–Barring interference, the old, with regard to phylogeny, leads the younger, in all states of consciousness.

Our first line of protection from danger is our instincts, followed by actions of the reptilian brain command, fight or flight, and lastly, the mammalian brain commands. In a situation where there are no interfering circumstances, the most primordial of these takes precedence over the least primordial. Thus for example, a person feels startled because a door jams unexpectedly, in an unexpected gust of wind, even though the individual may be in a safe environment, inside the home and in broad daylight, with no reason to suspect danger. In this instance, the three levels of consciousnesses, basic instincts (Permanent Attention), the reptilian and the mammalian were all alive and present but idling before the door jammed; the first to spring to action was the Permanent Attention, just executing the 'fight or flight' if the situation warranted it (the action of 'startle' is an attempted but instantly aborted flight); otherwise the rationalization of the mammalian brain then calms the situation and the individual realizes that it was only the wind that caused the door jam. If the door shuts loudly the second time the person is not startled, because now the mammalian brain is aware of the fact that there is no danger with the door shutting loudly. What it means is that but for the interference of the pre-information that occurred initially, the individual would be startled over and over again, each time turning around to be sure there was nothing amiss with the door jamming.

Another example is when a person stiffens as a thought passes though his mind, even though there is no present danger. This happened because, even though there was no danger lurking except in its imagination, it was enough to warrant the protective attentiveness to execute, and stiffening is a micro-mini display of an attempted flight, which is then aborted and calmed down by the cognitive mammalian brain making her realize there is no immediate danger. So also is the case of sleepwalking. Under normal circumstances, we would be sleepwalking always, but the release in REM atonia by neurotransmitters, during REM sleep, of norepinephrine, serotonin and histamine, (another form of interference), prevents us from sleepwalking each night, meaning the older brain would have assumed command, even though the younger mammalian brain was present, but for the interference of suppressing the neurotransmitters.

However, if conditions came back to a level field between the two, and the motor neuron movements were not suppressed, then the reptilian would come back in charge, and we would sleepwalk. This theory explains, for example why there must be boundaries, which a suspect cannot cross when defending actions involving sleepwalking; as it is not possible to include any actions that entails thinking or cognition in any action of sleepwalking since the reptilian is totally in charge at that period.

The Consciousness Recapitulation Hypothesis

From ontogeny to adulthood, homo sapiens recapitulated the phylogeny of consciousness.

The human, from embryo to adulthood recapitulated the phylogeny of consciousness. The human embryo grows from Vegetate (Plant) to Reptilian to Mammalian and then to self-consciousness which develops gradually until full self-consciousness is gained, sometime in young adulthood. It is the reason why human adults only recollect their childhood up to a point and never could remember what happened before that time. The period where recollection is impossible represents a transition from NDC to SDC and then gradually from SDC, self-consciousness is gained until young adulthood, when full self consciousness becomes apparent, and PDC is established.

The Regurgitation hypothesis of Consciousness

In consciousness, dissolution regurgitates phylogeny.

In the process of sleep, consciousness erodes and phylogeny regurgitates up to a point, for example in humans, the highest consciousness, to the reptilian consciousness. In the death process, the phylogeny of consciousness is regurgitated all the way unto death or zero consciousness. Therefore, as made apparent in the DF EQcons (Figure 1), a slowly dying individual will pass back through all the stages that organisms passed through in evolving. This explains situations such as the different stages of coma, in the process leading, from the highest level or maximum consciousness in a healthy human adult through the Mammalian, Reptilian and Plant (Vegetate) Consciousness to the lowest minimum, which is zero consciousness or death, in a continuum.

Sleep State Of Consciousness Decoded

Evolution of consciousness

'Life 'surrected' from death, which is zero consciousness, and evolved from there in a gradually increasing continuum, the Consciousness–Continuum' (CC), through different stages, broadly, the single–celled, plant, simple animals, complex animals and sea creatures, amphibians, reptiles, mammals, to the highest, which is human consciousness. In the order of:

Surrection→Sporadic Attention→Permanent Attention→Non-Discernible Concentration→Sporadic Discernible Concentration→ Permanent Discernible Concentration.

The 'DF Equilibrium of Consciousness' or the 'DF EQcons'

Life is consciousness in a delicate equilibrium, the direction of which is influenced by anyone or the resultant of all the internal and external stimuli effective on an organism, with phylogeny onwards in the forward aspect and backwards in the reverse.

- REM sleep when decoded technically is a syndrome, herein referred to as the DF Syndrome or DFS, in which we are compelled to re-enact our reptilian past, after having crossed into 3rd Class Reptilian consciousness (see Table 1); but which our biochemistry resists, in a condition known as REM atonia, or motor neuron movement paralysis).

- RBD (Rapid Eye Movement Behavioral Disorder) decoded is technically, just the observable aspects of DFS, bordering on and referred to as 'DFD (DF Disorder) manifestations.'

- Sleepwalking decoded results from the inactivation of DFS, culminating in a Disorder, herein referred to as 'DF Disorder' or DFD, in which without the muscle movement inhibition associated with the former, (or in which such inhibitions are over ridden), we actualize the reptilian re-enactment compulsion.

- The sleep cycle decoded is technically a cycle that balances our consciousness levels, herein referred to as the 'DF Cycle' or DFC.

According to the above equation, herein referred to as the 'DF Equilibrium of Consciousness', sleep time is one of a few times in the lifetime of a mammal in which the equilibrium goes in the reverse direction, and our phylogeny of consciousness regurgitates, which means theoretically, we could sleep back to death. Our consciousness as revealed in the arrows of 'Evolution of Consciousness' (above) faces the reverse side in our sleep. Therefore, sleep time is anything but the peaceful period it is perceived to be. In fact, as now revealed, it is a precarious period of mammalian existence; as the only time, apart from in sickness and the death process, in which the 'DFEQcons' spirals in the negative direction and requires the delicate balance maintained, as in an emergency resulting in the 'DFC'. As we sleep, our level of consciousness drops gradually but steadily and at a stage, it approaches a critical junction, at the uppermost edge of the so-called NREM cycle or the NREM + (Table 1), and if DF EQcons continues in the negative direction unchecked, then we could sleep into coma, since consciousness is a continuum. However billions of years of evolution meant that our biochemistry will not allow this, instead of which we are then compelled to get up to re-enact our reptilian past, in the DF Disorder.

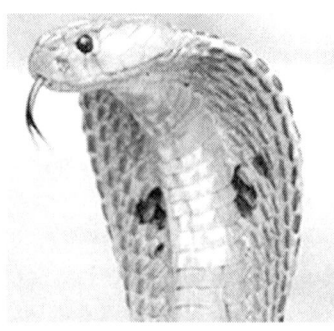

The Indian Cobra. Mammals can only sleep at the level of Reptilian consciousness

Mammals can only sleep at the Reptilian or Third Class consciousness state, as both the Mammalian and Human consciousness states are wakeful, sharp and focused states at which sleep is impossible; hence humans and mammals switch off their consciousness states and assume that of the reptiles, at sleep. Therefore, even though consciousness is a continuum that overlaps at the meeting points of one consciousness with the other,; no sooner are we asleep than we are in the reptilian consciousness state; and once in this territory, the natural thing is for us to want to get up to re-enact our reptilian evolutionary past, in a sleepwalking, glazed eyes world of our reptilian ancestors, which is the DF Disorder. However under normal circumstances this will not happen. This is because in the sleep cycle immediately preceding the NREM, or so called REM cycle, the release of the neurotransmittersnorepinephrine, serotonin, and histamine, responsible for motor neuron movements are completely suppressed, the condition called REM atonia, thereby paralyzing our muscles and rendering us incapable of getting up to do the reptilian re-enactment. Conversely, the unusual circumstance sometimes but very rarely occurs, in which either the release of the motor neurons are either not completely suppressed, or not suppressed at all. In the first case scenario, leg or other body movements while in bed may result; while in the latter case sleepwalking results; in which case the individual then open his/her eyes, gets up to walk around or perform routine activities, in a re-enactment of the level of consciousness found in reptiles. However even when there is complete suppression of the neurotransmitters, we still struggle against its effects, and try, albeit in vain to move our muscles or open our eyes and get up in a bid to re-enact our reptilian past.

The fluttering of the eyes in our futile attempt to open them is what is observed and referred to as REM or 'Rapid Eye Movement' which in fact is only a part of the DF

Syndrome. The so-called 'Restless Leg Syndrome' (RLS), is a 'mini-DFS' and will occur if the neurotransmitters are not completely suppressed. Our struggle against this compulsion serves to raise back our consciousness, not only stopping it from falling further into danger zone. but bringing it up to a safer level, the REM cycle, from where we yet again start drifting off into the sweet abyss of another cycle of deep sleep, NREM, and the whole process repeats itself over again, one cycle after the other, until the morning, culminating in what is called the sleep cycle, but which is technically nothing to do with sleep, but consciousness balancing, and herein referred to as DF Cycle.

Interestingly, both DF Syndrome and the DF Disorder serve the same purpose; which is to act as safety nets preventing us from sleeping too deep or into coma. The former, the first safety net, does so in a more subtle way and happens as a matter of course, because as we struggle against the paralysis of our muscles due to the suppression of the neurotransmitters, that struggle serves to wake us up a little bit (not completely) preventing us from sleeping into the danger zone (Table 1); while the latter occurring only when, for whatever reasons, the first safety net fails, in which case our compulsion to re-enact our reptilian past is actualized; which in turn serves to exercise us up raising our consciousness to safer levels, after which we then return to bed.

Sleeping Japanese Macaques.

Treating Sleepwalking? Deductions from the Theory

With what is now known about sleepwalking, it may now be possible to treat this condition at those rare instances when intervention becomes unavoidable, by using drugs that can completely suppress the motor neurons to bring about an artificially induced REM atonia; and cover up the systemic biochemical lapse that can bring about sleepwalking. Theoretically, maybe, but a word of caution. An artificially-induced REM atonia may succeed in ensuring that potential sleepwalking is prevented but at the peril of allowing NREM to continue deeper into the negative or reverse aspect of DF EQcons. Sleepwalking, embarrassing at it may seem, actually serve two purposes which are that by making the sleepwalker get up from bed and walk around for a while (Duration will depend on the level of NREM from which it commences, the deeper the NREM, the longer the duration), will in the first place stop any further ingress into an even deeper NREM, while at the same time raising the level of consciousness back to REM level, so that the DF cycle may continue as it ought to. Whereas, artificially induced REM atonia may end up only stopping the sleepwalking aspect but in the process pave the way for further ingress into the danger zone, by removing the safety net of the DF cycle; with the danger, of allowing sleep to continue beyond NREM into NREM+ and then coma or beyond. Tampering with sleepwalking unless absolutely unavoidable indeed puts the sleepwalker at risk as it may tamper with the DF cycle, important in maintaining the delicate balance of the DF EQcons. However, treatment of sleepwalking, where it couldn't just be left to run its natural cause, is possible, with the right approaches.

Dreams state of consciousness decoded

Dreams decoded is technically 'Partial Virtual Memory Recall'

As already made apparent by this theory, at DFS (REM), our consciousness level is brought back to near its highest in sleep; closer to wake up level than deep sleep. Our concentration level then, as shown in Table 1, is just enough to allow us to partially access memory, virtually, and we are able to do a 'Partial Virtual Memory Recall' (PVMR); much like imagining something in real life (wakeful hours), only that this is imagination at its wildest.

Imagination in real life is moderated by the cognitive abilities of our brain, whereas in imagination at the level of consciousness in which dream occurs, where only PVMR is possible, this moderating influence is absent. Therefore it follows that the closer to wake up level the dream occurs, the closer to reality it is and the more sense (the less wild) it will likely make, since the higher our consciousness level is, the higher the ability to access memory; and vise versa. Only this DF definition explains why for example it is possible to import and incorporate audible events around us, while still asleep, directly into the 'imaginations' we call Dreams e.g. phone ringing, birds tweeting etc." However, since whether at sleep time or wakeful hours, our basic instincts (Permanent Attention consciousness) which includes intuition is ever present (Table 1), it is possible for Partial Virtual Memory Recall (PVMR) to combine with our intuition and then logical dreams occur.

Furthermore, PVMR at a higher level of consciousness, or High PVMR can combine with strong intuition or other faculties of our 'Permanent Attention consciousness' to result in dreams that appear illogical or could not have been predicted using logical reasoning, or referred to as unpredictable but sensible and intuitive. In the consciousness state of dreams, ideas glimpsed from PVMR, in the absence of full cognition, interacts freely with our intuition and perceptive powers combining randomly to produce, at a high enough level of consciousness, different types dreams, from simply sensible to logical and illogical. At low levels of consciousness dreams are meaningless.

Therefore
Dreams   = Partial Virtual Memory Recall 'PVMR
                    (without the benefit of reasoning)
                    = Unreasonable Imaginations
Senseless Dreams = Low PVMR (wild imaginations)
Sensible Dreams   = High PVMR (almost as good as real
                    life imaginations) = Reasonable

Logical Dreams   =   High   PVMR   +   Intuition   (Logical Combination)

Intuitive Dreams   =   High   PVMR   +   Intuition   (Illogical Combination)

Sleepwalking (Somnambulism) state of Consciousness decoded

In accordance with another hypothesis emanating from this theory, which is that 'In consciousness, dissolution regurgitates phylogeny' (DelaninFadahunsy 2011), it means we only sleepwalk, as a matter of course, when we cross over from the mammalian consciousness, in sleep, to the reptilian consciousness, as we head in the wrong direction (dissolution) of the DF EQcons towards the 'Danger zone' (Table 1). Since in our phylogeny is the capability to live like a reptile, when we get to that consciousness level, the natural thing is that our biochemistry compels us to re-enact our reptilian past, if nothing stops us.i.e. no suppression of the motor neurons. When and if that happens and we get up to actualize that compulsion, all our actions, while in that level, mimic that of a reptile, which means we use our made-possible muscle movements only for the most mundane, ritualistic and casual of actions, reptile like. With what we now know therefore, sleepwalking itself is a non-insane automatism, occurring as a course of nature, whose function is to serve as the final safety net preventing sleeping mammals from slipping through the Consciousness–Continuum (CC) into coma or death.

All actions performed under sleepwalking can only be casual, ritualistic, or routine, involving no thinking or cognitive ability, whatsoever. The reptilian type of consciousness, or what is referred to as 'Non-Discernible Concentration' is non-cognitive, and simply ritualistic (Table 1). Therefore any action that cannot be classified as such cannot be performed while sleepwalking. A sleepwalker cannot even run, otherwise the sleeper would awaken, much less attack someone. Who ever heard of a sleep runner? Running, rare though, may not be theoretically impossible in, the only problem being that such a person will most likely wake up or collide with something and fall no sooner than s/he started running, whereas s/he could well manage that state of consciousness just walking. A reptile, mind you, could use that state to run or do any of its activities (usually mundane and procedural). The underlying problem that makes mammals such poor users of that state of consciousness being that we last used it over 350 million years ago, and will never compare with the reptiles which have been using it for that as an additional period, since our phylogeny began, and therefore are naturally excellent at handling a wake-sleep consciousness state, since it is their natural state of existence and the only state they ever knew.

Furthermore, and perhaps one of the greatest misconception of sleepwalking is that it entails dreaming. According to this new theory, nothing could be further from the truth. This theorist, like many others, is habitual with a DFD manifestation or so called Restless Leg Syndrome, but at no time was it because I was dreaming of something that entails use of legs (my wife once asked if I was playing some game in my dream). On the contrary whenever it happened, it was embarrassing and unknown even to me when awakened up, unless I was informed of such occurrences. Sleepwalking is technically no more than that, the only difference being that in DFD manifestation, muscle movement is partly restricted, whereas in sleepwalking it is totally unrestricted. Sleepwalking is absolutely incompatible with dreams level of consciousness in the light of what we now know, because a true sleepwalker never remembers a thing about what happened while sleepwalking. At the stage of sleep at which sleepwalking occurs, (Table 1), dreaming is impossible, since there is no access of any sort to memory. In any event only mundane acts of a reptile are possible in that state, were we in a position to live it out, since according to this theory, there is absolutely no access to cognitive memory, while in that state.

Perhaps what makes any claim of violent crime under sleepwalking most improbable is the fact borne out of another hypothesis of this theorist, which is that 'in the phylogeny of consciousness, the old leads the younger', barring any interference' (Delanin Fadahunsy 2011). Actions of aggression rightly fall under 'might makes right' reptilian, but when or just as the aggression is about to start, the primordial survival instincts kick in. This is because aggression brings on the full red alert in both the perpetrator and the recipient. It is associated in organisms' phylogeny with survival; since a duel's outcome is never entirely predictive, even when it seems clear that one party has an advantage. Therefore, aggressor and defender both switch to red alert, because, if not, an animal may win the contest but may sustain an injury in it that may eventually disable it or even lead to its death. Therefore aggression is taken very seriously indeed in phylogeny and used only when unavoidable. An attacking snake, say, is at its alert best to deliver the strike quickly and precisely and so is an attacking lion making a kill. None of these can be done while deeply asleep, by mammals, because it's not a state of best alert for them; the only state under which such actions are possible.

Therefore, even if we imagine a sleepwalker, for the sake of an argument, going into an attack; at the point at which the individual goes on the offensive, then switches to its highest alert the individual has to wake up because the primordial instinct of fight or flight has taken over. When that happens, the individual can no longer remain in the reptilian consciousness because that is not the human highest form of alert, and so must rise up back to the appropriate level of consciousness, which is human and hence awakens. Any organism involved in aggression whether attacking or defending does so at its most alert because it is associated with the survival instinct. Adrenalin is released at the onset of such aggression and that in itself will wake up any sleepwalker; since adrenalin and sleep are incompatible, especially deep sleep, which sleepwalking is. Therefore claims such as remembering nothing with such a rush of adrenalin is akin to a bungee jumper remembering nothing of the incident. When adrenaline levels rise heart rate increases, air passages are dilated and a fight-or-flight response is given; all of which are the exact anti-thesis of a deep sleep state such as sleepwalking.

The reptilian mimicry is effectively over for the sleepwalker, at the point at which an attack is about to start, and a struggle (entailing the switch on of the survival instincts) is about to begin; automatically turning on the survival instincts which means all potentials available to that individual must be deployed, which includes the use of the memory. But memory as made apparent in Table 1 is only accessible at mammalian or human consciousness, meaning the sleepwalker is already very much awake, since these states of consciousness are wakeful states only. An attacking animal is at its level best of consciousness because the survival instincts are turned on, so is the sleepwalker, which means he or she is back at human level full alert consciousness, even as the battle is about to begin. Therefore any claim that such aggressive act was perpetrated without self-awareness is nothing but ridiculous, for it means someone is fighting while deeply asleep. A snake fights at its full alertness, but because it belongs in a third class (Reptilian) consciousness, its full alertness is the same as its sleep; not so for a human being whose full alertness is first class (self)consciousness. Therefore, while a snake cannot wake up from its sleepy state, a human being can and will, if survival instincts take control which would happen if aggression comes in, particularly in the case of firearms. The sound of a gun wakes up someone sleeping instantly, because the survival instincts kick in initiating the fight or flight, regardless of whether the gun is silenced or not, since it is not a matter of the pitch but rather the association of such a sound with survival. Conversely, a phone may ring as loud as it can and may not wake someone up, until after some time, because such a noise is not associated with danger in our memory. It is different with the sound of a gun as the first shot is all it takes for the survival instinct to kick in. The kick-back from such a gun is another considerable factor as it will further increase the sense of danger. Furthermore, the act of getting bullets, cocking a gun in readiness, and then firing a shot are absolutely incompatible with actions attributable to sleepwalking, because they are not mundane, or casual acts, but ones associated in our memory with death and survival. Furthermore, the sequential nature of these actions is absolutely incompatible with a confused state of consciousness such as sleepwalking.

Loading a gun with the intention of using it to kill, apart from the fact that intent and sleepwalking are incompatible, are acts that will kick in our survival instincts because anything related to danger and death brings on the survival alarms, regardless of whether an attacker or a defender. These acts are not synonymous with sleepwalking acts such as walking (most preponderant act in sleepwalking), cleaning, sitting, standing etc. Yes, it is possible to dream that one is loading a gun and even firing it at something; but arguments such as these are no longer tenable, in the light of a full understanding of this theory, in defense of sleepwalking violence. This is because sleepwalking does not involve dreaming, at all, since according to this theory, we now know that we need a partial access to memory for dreams to occur; reason why dreams occur at close to wake up consciousness levels or REM cycle. Sleepwalking is acting out our reptilian phylogeny, in deep dreamless sleep, and a reptile even though in a world of sleep—wakefulness lives a ritualistic but real life and is not dreaming; but like an automated machine, is repeating what it knows well and what it has done through the millennia to survive and preserve. Sleepwalking or DFD occurs only in deep sleep and that is why unlike in dreams, no genuine sleepwalker ever remembers a thing about the incident; since according to this theory, there is absolutely no access to memory in that state of consciousness.

Furthermore and as should be expected, animals and humans are such poor users of the reptilian state of consciousness because they last used it, actively, over 350 million years ago ; and only now use it for the necessity of sleep. Therefore, a sleepwalker, unlike a reptile is confused, befuddled, uncoordinated and helpless in that state. She can only barely manage the most basic of tasks which must be very routine to her, and even at that, will perform very poorly. A sleepwalker may casually pick up a rag to clean and no sooner than she has dragged it over a table surface, the rag may drop off from her hands, in the utter confusion that the state entails. In the only self-sleepwalking episode remembered by this theorist (with a habit of Restless Leg Syndrome), I had woken up in the morning to find my pajamas top by the entrance to the bathroom whose door was unusually ajar. I was fastidious at locking the bathroom door, which leads directly to the bedroom, before going to bed. I also noticed that car keys and other items removed from the pajamas trousers, and kept by the side of the bed, before I slept, were in disarray. I knew I had slept walk. I have a habit of taking off my pajamas top and using it as some form of support for my body when about to sleep; but anytime I get up from bed, will almost invariably grab the top instinctively to put on before going anywhere. Therefore, I knew I had gotten up sometime that night, grabbed the pajamas top and headed towards the bathroom, opened the door, but didn't go in (nothing was disturbed in there and it wasn't used). After opening the door of the bathroom and probably fumbling there a while, I must have dropped the pajamas top and returned to bed. The items near the bed were in disarray because I did not side track them in my confused walk to the bathroom door (which I might have done, were I performing the actions while fully conscious); and it was possible I also sat up in bed for a while, like I often do, before actually grabbing that top and heading for the bathroom, resulting in the disarray of the items by the side of the bed, since they were where I would have put my legs if indeed I sat down. I also remember that earlier as a child, my sister of about 5 years (I was about 7) getting up from sleep at about 10.30 p.m. looking confused and glaze-eyed, walking past where I sat (listening to Mum and Dad talk; the entrance door still unlocked) and heading for the bushy playground about 70 yards from the house. I was young and didn't understand what she was doing, but got up to see where she was headed at that time of the night. She passed by so quietly, my parents didn't notice, engrossed in their conversation. She stopped once or twice briefly scratching her head and then continued. She was about fifteen to twenty meters from the house when I became alarmed and called their attention after realizing she wasn't trying to turn back. Dad went after her, guiding her back gently to the house and her bed and saying she was sleep walking. These are the uncoordinated, disjointed, confused, sort of actions possible in sleep walking, except for walking itself, which is the only act with which we appear comfortable in that state. A sleepwalker cannot be imagined to step on a cockroach to squash it because even that requires some level of concentration, coordination and intentionality; all of which are absent in that state of consciousness, much less go on a violent attack directed specifically at someone else.

Perhaps more than anything else, humans like other bio-chemical organisms are, under natural circumstances, nothing but a compilation of our phylogeny, therefore if nothing in the latter suggests that an attack, especially to kill, is possible for no obvious reason, and much more importantly, if it serves no obvious or deductible purpose, then aggression under the guise of sleepwalking cannot be justified. Moreover, animals are affectionate, and none will attack a loved one, without reasons under any state of consciousness. A lion will not get up to attack and kill its female lion, purposelessly, for example, neither will it kill its own cub; nor will a reptile.

There have been instances of 'bizarre' killings observed in nature, but a closer scrutiny will always reveal an underlying reason. African lions are known to kill cubs if such were inherited from another male, the purpose being to enable the new male mate with the female of the pride and produce offspring of its own, before it too is chased away by a stronger male taking over the pride. Just as some mammals e.g. rats are known to practice infanticide, but always for one deductible reason or another, invariably not unconnected with survival.

It is pertinent to say that whatever cannot be done while awake, will not be done while sleepwalking because the latter only encompasses innocent ritualistic actions, done while deeply asleep, with absolutely no awareness. Moreover, this theory makes clear that sleepwalking, just like mini-DFS are not deeds of acting out a dream but rather involuntary actions, serving the purpose of raising our consciousness level. Sleepwalking is phylogeny regurgitating, as the sleeper heads down the wrong side of the DF EQcons. A sleepwalker is just like an automated machine performing only what he or she knows well (because his brain is not involved in any such action) in a reptilian consciousness, in which his/her ancestors once lived.

The Oddity Hypothesis (Oh)

The troubling issue of sleepwalking violence, which can now be laid to rest with this theory, has made inevitable, if only for future reference, the 'Oddity Hypothesis' which states that 'Any action of an organism that is neither observable in nor deductible, inferable or traceable from phylogeny can only be considered unnatural or insane' (Delanin Fadahunsy 2011). Even though the hypothesis emanates from this theory, it is generally applicable. This is because even disregarding all considerations of consciousness and its states, and the fact that sleepwalking has not been much observed, nor have there been enough attempts to, in animals or other earthlings, with whom humans share a common phylogeny; however it is traceable directly to sleep, a common phenomenon, at least in mammals, and is therefore inferably natural. However, purposeless aggression, under which sleepwalking violence fall, is neither deductible, traceable, nor inferable from the phylogeny of earthlings and therefore, can only, at best, be considered unnatural or insane. There can be no other definition.

A snake will never attack unless provoked or going after a prey item, or defending territory, or for other survival-enhancing reasons, which do not include unprovoked, purposeless, aggression. Therefore to suggest that a snake will attack or kill another snake showing no aggression nor trespassing its territory, nor giving any cause for such an action, is highly unlikely. Therefore, aggression even in 'might makes right is not without purpose'. In the same vein, to suggest that someone would commit an act of unprovoked aggression against someone else, while sleep walking amounts to and belong in the ignorance of what sleep walking really is. It is akin to saying someone with a DFD manifestation (Restless Leg Syndrome), violently assaulted someone while in the act, on the bed. If this is impossible by any stretch of the imagination, then so is criminal violence while sleepwalking; because both are technically the same thing and performing the same purpose, which is to wake the sleeper up a bit (not completely) so he does not sleep into the danger zone or coma. Aggression, even in 'might makes right', is not without purpose. In fact it is uncommon to find examples in organisms living in their natural environments where aggression is totally unprovoked or absolutely purposeless. Otherwise, snakes would go looking for other snakes or animals for other animals or humans to kill them randomly. Only an alien organism will do something that is not in the phylogeny of earthlings. *As such, violent aggression and sleepwalking are incompatible.*

This is because our consciousness is for a purpose, which is to counter our ultimate fate. All earthlings share this common purpose and this destiny, which are survival and death respectively, since death is a no-consciousness or zero consciousness state. Therefore all actions of earthlings, from the tiniest to the biggest and the wildest to the meekest are directed towards fulfilling this purpose. Therefore earthlings are very purposeful and as such there is not one single action that is performed by them that is not directed towards the fulfillment of this purpose. Attacks by wild animals on their hapless preys, snake strikes, and all forms of aggression by them only serves to fulfill that intention; making sleep walking aggression stand out like a sore thumb, as the only purposeless aggression known, perhaps in the entire 4.5 billion years of our evolutionary history. Therefore sleepwalking violence could only be described as either an unnatural act, which means it is calculated and intentional and not borne out of a natural cause of action; or it is due to insanity.

The Legal perspective
and mens rea*

In the light of this theory therefore, since sleepwalking is now deciphered, with its only function being to serve as the safety net preventing sleeping mammals from the danger of coma or death, it means the act itself is technically a naturally occurring non-insane automatism. The use of the sleepwalker's body, muscle movement and entire biochemistry are as such directed, pre-programmed by nature, specifically, at performing that function; and could therefore, no longer be redirected or re-assigned for any other simultaneous use (such as aggression), anymore than a sneezing nose could be redirected or re-assigned to smell a flower, at the exact instance of the sneeze, simultaneously. In other words, it is not possible to be in a state of sleepwalking, non-insane automatism, while at the exact same time performing yet another purported act of non-insane automatism, such as aggression in sleepwalking was argued to be. This is because, with what we now know, only the non-insane automatism of sleepwalking pre-programmed as it were by nature, like other automatism acts, can take place in sleepwalking.

Lord Denning's dictum in the leading case of Bratty v. Attorney-General for Northern Ireland (1963) AC 386, at 409 defines automatism as:

> '... *an act which is done by the muscles without any control by the mind such as a spasm, a reflex action or a convulsion; or an act done by a person who is not conscious of what he is doing ....*'

Therefore, once the action of automatism has started it has to run its course. Sneezing and epilepsy, for example, are naturally occurring acts of non-insane automatism, much like sleepwalking. Sleepwalking will run the course defined strictly by the function it has been pre-set by nature to perform, which is to 'exercise' a sleeper up a little (raise the level of consciousness), so s/he does not sleep too deep into the danger of coma or death. Therefore no other action is possible while sleepwalking is running its course, just as the nose cannot be used for any other action at an instance of sneezing, because the act itself is the non-insane automatism, neither can an epileptic use the equally uncontrollable actions of a seizure to commit a murder, at the same instance. Consequently, as already submitted, since no other action is possible during an automatism act, any other act including aggression is impossible during the act of sleepwalking. Consequently any such other act can only be considered as being due to other factors such as may subsequently establish a *mens rea* or the possibility of insane automatism.

Sleepwalking is here defined as an non-insane automatism exercise act whose function is to raise a sleepwalker's consciousness level, so s/he does not fall through the consciousness-continuum to unconsciousness or coma.

In addition, since it has unambiguously define the function of sleepwalking, this theory now makes possible the delineation of the limit above which an action could be deemed impossible under sleepwalking, one that will be more precise than fluid descriptions such as actions that are casual, ritualistic, procedural, and non cognitive. We now know that sleepwalking functions only, as far as, to awaken the sleepwalker a little, but not fully, and then to return such individual to sleep, still not awoken. Therefore nature has made these activities mundane and as gentle as possible to accomplish the given objective of exercising up, without waking up the sleepwalker.

Therefore, an action will be deemed sleepwalking incompatible (*swinc*) if it is deemed in excess of (>) the minimum (x) required to arouse an individual up from deep natural sleep (devoid of external influence, such as alcohol, drugs, etc), and sleep walking compatible (*swac*), if it is deemed below or equal to x; since the actual maximum stimuli effected by nature itself, on the sleepwalker during sleepwalking, is one that is just enough to raise the individual's consciousness level, but still well under (<) the minimum stimuli required (x) to arouse her/him.

This formulation will still leave a generous allowance of stimuli (force) exert—able by the sleepwalker, which nature itself will neither have exerted on nor permitted the sleepwalker to exert in the non-insane automatism exercise to wake him up; when juxtaposed with the fact that while in that state, nature will neither exert nor permit him/her to use as much force as might arouse him/her in any activity/ ies s/he may chose to engage in.

Therefore:

An action = *swinc* (if it is > x)

and

An action = *swac* (if it is not greater than x—i.e. it could be less or equal to x)

where

x = is the 'wake up force' or the considered reasonable minimum stimulus (force) required to arouse an individual up from deep natural (uninfluenced by external factors, such as alcohol, drugs etc.) sleep.

In other words, a sleepwalker may use as much force, at any given instance, in any activity of choice while sleepwalking as may be reasonably considered able to arouse a deep sleeper from sleep and no more for her/his actions to be *swac*, as anything in excess of that will have to be considered a *swinc* action.

The concept of this formulation, in the light of what we now know about sleepwalking, is based on the common sense presumption that an individual in normal sleep (not unconscious nor in concussion or coma) will wake up when a certain minimum amount of force (x) is applied on him/her and in the same vein, it is reasonable to assume that s/he will be aroused if under any circumstances, s/he were to apply a force equal to x on anything while in that state of non-insane automatism act, imposed by nature. This in turn only tends to corroborate the casual, non-cognitive, mundane, procedural, ritualistic descriptions of the nature of sleepwalking acts.

Even from the above analysis alone, it is apparent that the force required in virtually all known cases of physical assault reported in sleepwalking, is greater (virtually all ridiculously so) than x, which will tend to make practically any type of aggression, save perhaps, for instance, inadvertently nudging someone, in sleepwalking *swincs,* while in sleepwalking murders (to take out a human life in a life/death survival struggle), the order of greatness from x, is such that it would put the action in an entirely different class of *swincs.*

It is also noteworthy that while nature decides the amount of force a sleepwalker may use in any activity of choice engaged in, it has placed no such restriction on the type of activity/ies, so long as the delicate definition of 'not forceful enough to arouse him/her' is strictly adhered to. However, common sense will tend to dictate that it will have to be activities with which the sleepwalker is very comfortable at wakeful hours, since all of it will be performed in an unconscious state. This is because it is quite appropriate to expect that when our biochemistry is given a free hand, by nature, to choose an activity that would not wake us up while sleepwalking, it will choose one with which we are most comfortable. It will also have to be activities that such individual can perform subconsciously while awake, i.e. activities possible without even giving a thought to what they were doing, which can only be those with which the individual is very familiar or one that is fairly leisurely, such as walking.

The myth of driving while sleepwalking.

The above explanation makes obvious the fact that driving is almost totally impossible when sleepwalking. Perhaps, the farthest a sleepwalker will get is to open the car door, sit and slip in the key into the ignition, and possibly manage to start the car. It is impossible for a sleepwalker to actually do the driving. This is because driving itself involves concentration, which is not possible in a deep sleep state such as sleepwalking, without the sleepwalker waking up or crashing the vehicle, almost as soon as it moved. Driving over a distance by a sleepwalker belongs in the myths of our past ignorance of that phenomenon. The concentration act of looking ahead and steering when driving is a stimulus that will wake up a sleepwalker and as such, driving is 'swinc'. We all know the danger of driving while feeling sleepy as a momentary loss of concentration is all required to have an accident. Therefore to assert that someone in deep sleep state of consciousness, such as sleepwalking is, moved a car one foot without crashing, as soon as it moves, belongs in the ignorance of the past on the phenomenon.

Sleep as a necessity

This theory made clear that once life evolved from sedentary to mobile life, sleep became a necessity. Organisms realize that the night offers dangers from predators and therefore gather food or prey on organisms, mostly, during the daytime so as to hide away at night, rest and conserve energy for the next day's sojourn, They soon recognize how restored and ready for that day they were when they have slept and as such came to appreciate the importance of the restoration that sleep brings. These are facts that have now being corroborated by studies which have found out that after the initial onset of the first delta activity during sleep, certain growth hormones secreted are essential for growth in animals' infants, as well as being very crucial in repairing body tissues (Moorcroft, 1993). Moreover, a biochemical system such as organisms are, needs a chance intermittently to repair the wear and tear caused by the daily activities, such as going after prey items, food gathering, fight or flight executions etc and sleep time offers the best opportunity for this. Perhaps nothing better demonstrates the importance of sleep than what occurs when there is sleep deprivation. In a study with human subjects Dinges and Kribbs discovered that performance on short tasks is not impaired when individuals are sleep deprived; however, performance on longer tasks which require sustained attention becomes impaired. In other experiments subjects report perceptual distortions and or even hallucinations (Franken, 1994). Rechtschaffen's study on rats deprived of sleep for between 5 to 33 days showed harsh effects. In the research, the rats began to look sick and stopped grooming themselves. They also became frail and clumsy. Some of them died and some had to be sacrificed. On autopsy, signs of acute stress were observed in form of enlarged adrenal glands, stomach ulcers, and fluid in lungs (Carlson, 1991).

Waking up a sleepwalker

This theory has now made it evident as explained above that it is unwise to wake up a sleepwalker, and it is preferable to let the act run its course and the sleepwalker naturally returning to sleep by themselves, no matter how long, so as to let the DF Cycle reset with time, as intended by nature. While effort could be made to be with them anywhere they go during the act, there need be no fear that s/he may walk and trip, if within a very familiar environment. This is because in such a situation wherever a sleepwalker trips while sleepwalking, the individual will most likely trip at wakeful hours, because even though deeply asleep, s/he could move around and act freely, like, say, a snake, in a well-known environment.

Why Children are more predisposed to sleepwalking

More children sleepwalk because children are only able to use two states of consciousness, the reptilian and the mammalian well. This is because unlike adults, who in addition use the human consciousness very well, children are only just developing a database of self and environment, which forms the larger picture of the 'picture-within-a-picture' self-consciousness capability, and until well formed, they will tend to be more predisposed to the use of their mammalian and reptilian consciousness. Therefore lapsing back to the latter state more frequently is not unnatural and it's the reason why they tend to sleepwalk more, as they swing from mammalian to reptilian consciousness at sleep. As will be advised by most sleep clinics, it is important to take simple precautions, such as locking doors, especially entrance doors. In another forum, other perspectives, hitherto unexplored will be discussed whereby sleepwalking could be halted without tampering with the DF Cycle.

The Consciousness Compulsion, (Concom) Hypothesis

Quite apart from the biochemistry of evolution of life, exists the Survival or Consciousness Compulsion (Concom), without which perfect biochemical possibilities regardless, conscious life was impossible. It is the inertial breaking force essential for the evolution of conscious life, and which later became, in its step-down form, the 'will for survival', which is just as equally essential for its continued existence and without which it will easier revert to the natural state of sleep. Sleep, it would appear, is our natural or restful state of being, which needed to be bridged at origin, sleep abysmal or zero consciousness, before consciousness was gained. Therefore, in order for consciousness to be gained, there was the need for the compulsive force, Concom, to scale that barrier, for to suggest otherwise will break the laws of inertia. Therefore, in the light of this new theory, any consideration of evolution of life, without consideration of the Concom factor could only be considering unconscious life, such as some form of robotic living creatures, which is quite possible, in some 'zombie' form of existence that also evolved by self-replication; or why not, if not? Conscious life implies, as now revealed by this theory, a purpose for existence, which later becomes the will to survive. It is the reason why robots will forever remained programmed and never be made truly conscious, unless a way is found to imbue them with Concom.

This is apparent even from the mere observation that even though only humans are self-conscious, amazingly all living things, even the so called simplest life forms, that occupy the lowest stratum of the CC, want badly to remain conscious, to live, even though they do not know why. They go to great lengths to keep living anyway, when it might have been natural and expected of them, not to really care, either way. Why do they care? What does it matter to them and why does it matter so much, just the way it matters to us in a life or death struggle. The most plausible explanation will be the existence of the Concom or the 'Survival will'; after all, robots, even the most complex and sophisticated forms, don't care if they live or die, even though they would appear to be far more superior in intelligence than natural simple life forms. Therefore Concom is conscious life's 'propulsion engine' without which it is not conscious life, and life without it, even though a distinct possibility would have been purposeless, with neither the care nor will to survive.

Asimo; a bipedal robot

Furthermore, while reasons could be adduced for the necessity of consciousness and its gradual increase once life had evolved, no such thing could be said of where and how—how it acquired the force necessary to lift up from its state of zero consciousness inertia of sleep-abysmal.

Consciousness is a biochemical concept, unfeasible as an electromagnetic or such other concept due to the impossibility of imbuing an object, say a robot, with the Consciousness Compulsion, without which it can only essentially remain programmed, one way or another, but never really be conscious.

A brief word on Julian Jayne's book 'The origin of consciousness in the breakdown of the bicameral mind'

Away from the jubilee of the erudition of an author like Julian Jayne, and when there is the readiness and open mindedness to look at the issue more critically, we have got little choice, especially in the light of this theory, but to listen to cerebrals such as Ned Block; who have argued, that the eminent author had confused the emergence of consciousness with the emergence of the concept of consciousness. Such arguments, now corroborated by the theory, elucidate the fact that humans had been conscious all along, but did not have consciousness as a concept; reason why they never discussed it in their texts.

As already discussed here before, the nature of self-consciousness itself is such that even when we have gained it, we are unaware and long after we have started engaging its tremendous potentials, we are still totally oblivious of that fact. Suffice to say, according to the tenets of this theory, that had we not gained self-consciousness when we did, approximately 2.588 Million years ago, which was a far cry from the period Jayne suggested (approximately 3000 years ago), there may never have been humans on earth; since we would most likely have gone extinct ultimately from the predation pressures exerted, especially on our ancestral line of apes, by the then *de facto* rulers of our planet, the formidable cat class. This is because, on a plane level field, in terms of raw power, cunningness and agility the cat class remained the fittest and would most definitely have outlived us, winning the blue planet for its class anytime. In order to win, we needed brain power, but the brain as already fully explained by this theory, is almost totally useless without a permanent presence of mind or self–consciousness, without which there could be no control of fire. Therefore it was the latter, more than anything else that won us the prized jewel. As already explained, the nature of self–consciousness is such that long after the Hominins had put the cat class in the cage, they still were living in the oblivion of the fact that the lion king now belonged in folklores and in its stead, a new eternal ruler of planet earth, the *Homo Sapiens*, was crowned.

Concerning perceptions

Since we perceive with SDC, our feelings and emotions towards our perceptions arise primarily out of and as a result of our comparisons with PDC, which is ourselves and are therefore essential products of our unique ability for self-reflection; an ability not shared by other mammals who only react to such perceptions in terms of what they represent, e.g. dangers, threats etc. Therefore the more different a phenomenon is from what we are or what we know, the more emotionally rewarding or otherwise it can be.

The Mirror Test revisited in the light of the conceptualized 'DF Continuum'

The so called 'Mirror Test' for animal consciousness is flawed in that it was constructed to assume a scenario that was less than likely, but which would have suited a different scenario that would then have contradicted the inference. Consider these two scenarios side by side.

SCENARIO 1

Animal looking at itself in the mirror feels:

Oh! Is that my reflection in the mirror?

Then accesses memory (RAM) to compare what it knows of itself (virtually everything there is to know) and the image in the mirror and feels:

Oh! Yes, it must be me!

Oh, hang on. There's a spot on me in the mirror. Where is it on me?

And then goes ahead to search for and then located the spot on its body.

SCENARIO 2

Animal looking at itself in the mirror feels:

Oh! That's another of my mates (species) in there. It seems okay (not in any obvious distress), so what is doing there? How did it get in there?

However, while everything might seem okay at a casual glance, something about the mirror image isn't just right.

Then accesses memory to compare what it knows of its species (virtually everything there is to know) with the image in the mirror and feels:

Oh! What's that spot on the mate? If my memory serves me correctly, we don't have such spots! How did it get that spot on itself?

Hang on! Do I have it too?

And then goes ahead to search and then located the spot on its body. Finding it and knowing from memory that they don't have such spots; it is smart enough to know it has possibly rubbed on its body from somewhere. Several times in the past; it's probably gotten off items from its body or rubbed off dirt or stains from its feather that it knows does not belong there. Furthermore, if for whatever reasons its species have just started growing some strange spots, it wants to get out to compare and confirm positively and then file it away in memory for future use, as even survival may someday depend on such innocuous knowledge.

Comparing the above scenarios, it is obvious that Scenario 1 is very unlikely and the test is flawed primarily because, as far as we know of it, the animals in question have never been told of the existence of a mirror, let alone experiment, and familiarize with it. Therefore the first feeling of *'Oh! Yes it must be me!'* is highly improbable. To feel like that, the animal must, at least, be aware, or must have previously stored in its memory, that such a contraption, that can reflect an image exists. If the memory is blank of any such notion, then it is only seeing another of its species in the mirror. However it is not impossible for it to assume and understand that the image in the mirror is its, if for example it had been previously exposed to and allowed to play around the mirror long enough (including going behind it to satisfy its curiosity) and been given the chance to explore or get used to what mirrors do; in order to file away in memory that the contraption can do such things as reflect an image and that there is no other thing behind or inside the mirror. Therefore self recognition in the mirror is not impossible for animals, only probably not in the way it was assumed in the test.

In any event, even if did pass this test, it would not prove self-consciousness, by the logic of this theory. While self recognition is possible in Sporadic Concentration Consciousness; as has already been explained in this theory, it does not equate self-consciousness. A calibrated table of 'Degree and Type of Consciousness' is the future concept of this theory. The methodology will involve correlating the consciousness/cognition level of each and every organism to that of the calibrated levels of the development of the human, Homo sapiens, from conception (or baby) to adulthood; the latter being the only life form that transverses all forms of consciousnesses, as its ontogeny continuing to adulthood recapitulated the phylogeny of consciousness. For example neo–piagetian stages applied to the maximum stage attained by various animals have shown that spiders can attain the circular sensory stage, coordinating and perceptions, while pigeons attain the sensory motor stage, forming concepts (Wikipedia Online, 2011); so spiders and pigeons can now be scaled on the 'DF Continuum', which is the conceptualized Consciousness scale.

Conclusion as Positional Synopsis (Caps)

INTUITION AND INTELLIGENCE

INTUITION

Intuition has been considered through the times and defined variously, including as below:

'Intuition is a combination of historical (empirical) data, deep and heightened observation and an ability to cut through the thickness of surface reality. Intuition is like a slow motion machine that captures data instantaneously and hits you like a ton of bricks. Intuition is a knowing, a sensing that is beyond the conscious understanding—a gut feeling. Intuition is not pseudo-science.'

– Abella Arthur

'Intuition (is) perception via the unconscious' Carl Gustav Jung

Carl Jung in his theory of the 'ego' further described in 1921 in *Psychological Type*s, intuition was an 'irrational function' opposed most directly by sensation and opposed less strongly by the 'rational functions' of thinking and feeling'

'Intuition may be defined as understanding or knowing without conscious recourse to thought, observation or reason. Some see this unmediated process as somehow mystical while others describe intuition as being a response to unconscious cues or implicitly apprehended prior learning.'

– Dr. Jason Gallate & Ms Shannan Keen BA

Rudolph Steiner postulated that intuition is the third of three stages of higher knowledge, coming after imagination and inspiration, and is characterized by a state of immediate and complete experience of, or even union with, the object of knowledge without loss of the subject's individual ego

Intuition is often discussed in spiritual context, and described or regarded as 'a conscious commonality between earthly knowledge and the higher spiritual knowledge and appears as flashes of illumination.(Wikipedia Encyclopedia: Intuition 2010)

One definition seems to be a common denominator, which is that it is almost generally agreed that intuition cannot be judged by logical reasoning.

The oddity hypothesis (oh) approach
Researching with the oh paradigm, which has been explained earlier in this theory and which infers that

any attribute of an organism is deductible, inferable, observable or traceable from phylogeny or otherwise it is unnatural or insane, intuition has thus been identified and here defined as:

An innate mechanism, in a cognitive organism, prompting automatic computation of the best instantaneous solution, whose quality is not time dependent, to a problem, using the organism's memory (mind), and perception, much like an electronic calculator computes instant solutions, without any thought process involved (Delanin Fadahunsy 2011).

The mind is defined by this theory as an organism's entire database of information, including all about self and its world, bio – chemically stored. Therefore, animals like humans have minds of their own, but while humans are more predisposed to engaging theirs through a thought mechanism called rationalization, animals, having no such level of intelligence, engage theirs through a mechanism called intuition.

This theory has identified therefore that intuition is a different kind of intelligence in its own right, here referred to as Intuition Intelligence (II), as distinct from Reasoning Intelligence (RI) of humans.

INTELLIGENCE

Intelligence, a word that is a derivative of the Latin verb *intelligere* (inter—legere) meaning to 'pick out' or discern, has been defined variously by different researchers, some of which, randomly selected, are as below:

Researcher	Quotation
Alfred Binnet	Judgment, otherwise called "good sense," "practical sense," "initiative," the faculty of adapting one's self to circumstances . . . auto-critique.
David Wechsler	The aggregate or global capacity of the individual to act purposefully, to think rationally, and to deal effectively with his environment.
Lloyd Humphreys	". . . the resultant of the process of acquiring, storing in memory, retrieving, combining, comparing, and using in new contexts information and conceptual skills.
Cyrill Butt	Innate general cognitive ability
Howard Gardner	To my mind, a human intellectual competence must entail a set of skills of problem solving—enabling the individual to resolve genuine problems or difficulties that he or she encounters and, when appropriate, to create an effective product—and must also entail the potential for finding or creating problems—and thereby laying the groundwork for the acquisition of new knowledge.
Linda Gottfredson	The ability to deal with cognitive complexity.
Stenberg & Salter	Goal—directed adaptive behavior.

Reuven Feuerstein	The theory of Structural Cognitive Modifiability describes intelligence as "the unique propensity of human beings to change or modify the structure of their cognitive functioning to adapt to the changing demands of a life situation

(Source: Wikipedia Encyclopedia online)

The origin of intelligence

Intelligence, like consciousness, almost certainly originated with life itself, right at the origin of living cells. Organisms discovered very early in time that competition for scarce resources meant they needed to device means not only to get those vital resources, but to also keep competitors away from snatching such resources from them. Furthermore, they had to device means to be able to, for example discern competitors and foes, who may be camouflaged, from family and friends. Moreover, fierce competition meant organisms not only had to devise and discern, they had to learn to keep improving on these faculties in order to keep ahead of their smart competitors and fierce foes or rest on their oars and allow themselves to be outsmarted and exterminated. Additionally, survival at the cellular level meant the more potent the toxins secreted by organisms for attacking and defending purposes, the better their survival chances, individually or collectively. This meant that organisms were actually already masterminding their situation and each other at as sophisticated a level of struggle as virtually amounts to biochemical warfare. Activities such as just described exist only in intelligent organisms, and the capacity for such activities is what is intelligence. Therefore it is only logical to assume that no sooner had life begun than the use of intelligence was already firmly established in earthlings. As such, there hardly can be any gainsaying the fact that earthlings have always been an intelligent lot. And since intelligence, as already explained involves masterminding or being on top of situation, it is only rational to assume that intelligence cannot exist without a mind, where mind as explained earlier in the theory is the database of total information available to an organism, stored biochemically. This is because in order to make progress from any given point, position or situation, an intelligent life must have an idea of that present point, position or situation to be able to attempt to device the means to achieve the desired progress, as anything short of that could only either mean retrogression or at best going round in a circle which means effectively remaining static. Yet, there is hardly a doubt about the fact that earthlings, as explained before, imbued with concom, are instilled with trying to make progress only and will therefore neither naturally retrogress nor remain static. The only conclusion that may be drawn from these facts is that all earthlings have minds of their own, where information is intrinsically biochemically stored about, at least, past knowledge and the present position, from where they can then initiate solutions, consciousness compelled (concompelled) as they are to make progress. Therefore if intelligence is the art of devising means, then the mind is the obligatory tool, without which there can be no devising. The inception of intelligence at the origin of living cells took the same curve as explained earlier, in the theory of the development of consciousness, increasing in complexity with passing time and ultimately culminating in the highest, human intelligence.

Intelligence is hereby defined generally as the art of devising means and specifically as any innate physiological or other such natural mechanism or process, involving the use of the mind, imbuing the capacity to initiate solutions to problems, internal or external, in a living organism; while the mind is defined as the database of the entire information potential, intrinsically, biochemically, stored in an organism. In higher intelligence, part of such storage is centralized in the brain.

The Four Tier Intelligence (FTI) hypothesis

Therefore this theory postulates that there are four types of intelligence which are:

1. *Cellular Intelligence (CI)*
   CI is basic intelligence common to all life forms. It operates without cognition, using physiological or other such sensors to monitor, react and initiate solutions to perceived problems. Example of cellular intelligence in higher consciousnesses are our instinctive abilities (a hand inadvertently brought close to fire is withdrawn using our primordial

CI), so also is the immune system in mammals or other organisms, which initiates solutions to internal threats. The art of devising means to protect our system from enemy organisms, is made possible only by the carry over from the cellular 'biochemical warfare' aforementioned, at the origin of living cells and not a means devised by our reasoning intelligence, since we are not even aware of what is happening at that level, unless and until that intelligence gets outsmarted by invaders, and we fall ill. There are various other displays of such intelligence in all other life forms and are especially dramatized in plants due to their mostly sedentary mode of existence e.g. actions of insectivorous plants, photosensitivity, plants' other sensitivities, communication, etc.

2. *Procedural intelligence (PI)*

PI Involves no cognition and only operates by innate routines, practices and procedures learnt over million of years of evolution, e.g. as in reptiles, animals, humans. Sleepwalking, for example, which is acts of performing normal daily routines while in a state of deep sleep, involving no cognition or thought process, is a feat made possible by the vestiges of PI in mammals. Animals are able to play games and get into fairly complex rhymes with human counterparts, using their PI only.

3. *Intuition Intelligence (II)*

II Involves cognition and the use of memory but no reasoning capability, in which the best solution, whose quality is time independent, to a problem is automatically generated instantaneously through computation of available data and comparison with other available options, in a process called intuition, using information from the mind (memory), and perception, where the latter could be sensory or sub – sensory, examples of which are animals and humans. II may be referred to as animal intelligence, since the faculties of this level of intelligence are almost non – existent in humans, having been badly eroded due to total reliance on the exponentially superior Reasoning Intelligence (RI). So called intuition and intuitive dreams, discussed earlier in the theory are good examples of the vestiges of II in humans. Another example is talent in humans, which is a natural aptitude, not attributable to Reasoning Intelligence, depicting an excellence in the art of devising a particular means, which may then be expressed with or without the assistance of our Reasoning Intelligence.

4. *Reasoning Intelligence (RI)*

RI involves cognition and the use of reasoning abilities, in which solutions, whose qualities are time dependent, are arrived at through the calculation and consideration of the options available, using the mind, in a process called rationalization e.g. as in humans only. RI may be referred to as human intelligence, as reasoning is possible only with self consciousness, which as earlier discussed, only humans are capable of.

Sensory perception is here defined as those experienced through our normal five mammalian senses, while sub-sensory perceptions are those more 'felt' than perceived, including those imagined, as in intuitive dreams (HPVMR), explained earlier in the theory. An example of what is meant by sub-sensory perception, apart from dreams or imagination, is here described:

A partial perception may occur in which it appeared there was a movement seen out of the corner of the eye and at about the same time an almost imperceptible rustle or some noise that could be simply filtered out by our rationalizing intelligence, RI, as part of background noises, caused by the wind or something. However, while the two occurrences might have in fact meant nothing, the fact that they occurred in tandem, which was lost on the RI, was not lost on the vestige II, which then initiates the intuition process, automatically searching the database of information we call mind to match with similar past experiences and comes up with an instant solution, which is that those two happenings occurring as they did concurrently was not a mere coincidence and could only mean one thing. At which time, one is alerted and suddenly becomes more aware of the environment, with a feeling of, say, not being alone or being followed, where before, one had imagined being totally on his/her own.

Perhaps, the greatest distinguishing factor between II and RI is the absence and presence of reasoning respectively and it is this that has set the human race apart from animals, more so than any consideration of the capabilities brought on by our exponential level of intelligence. This is because if animals could reason, it is not impossible to rationalize that they could live a settled and civilized life much like humans, in spite of their lower level of intelligence. Even without any reasoning ability on their part, the amazing architectural feats by colonies of bees and ants, for example, and their level of organization, including hierarchical and social, will justify such rationalization. Reasoning, which could be by common sense (a term we use when we mean low level intelligence) does not necessarily always need to involve a high level of intelligence such as required for mathematical problem solving for example, and yet even though they are incapable of reasoning, animals are quite capable of these rather abstract levels of intelligence.

Thus while for example, 'young chimpanzees have been known to outperform human college students in tasks requiring remembering numbers and pigeons have been shown to outperform humans on the Monty Hall Problem, a probability puzzle' (Source' 'Animal Cognition' Wikipedia Online Encyclopedia 2011), the rationalization levels of these animals are next to nil. Their inability to rationalize cannot therefore be attributed solely to their level of intelligence. Yet, there is hardly any doubt about the fact that it is this inability to reason that makes them animals.

The absence of reasoning is why animals are wrongfully accused of being unpredictable. The reason for lack of reasoning as earlier posited by this theory is due to a lack of permanent presence of mind or PDC. Therefore while human consciousness rely on the ability of our superior brain, with PDC serving as a permanent background, allowing for a seamless integration of thoughts even when such thoughts are disjointed; no such thing exists in animal consciousness, since there is no PDC. Thus, whenever there is the need for a seamless integration of actions as, for example, with animals performing routines in a circus or bottlenose dolphins in rhyme with human counterparts, these animals lapse back into the primordial PI intelligence, which is tailor made for routines, whereas the humans are upscale, using the sophisticated RI, for the same routine.

Therefore, while if anything should change in the routine, such as a missed cue, the humans are quite capable of making up for it to continue the rhyme, the animals are unable to, and may even lose the entire cue altogether, even when such missing cue may necessitate nothing near the level of intelligence required at performing mathematical puzzles as earlier discussed, at which they could outperform human college students, for example. The reason for this is that they cannot rationalize or reason out a situation, like humans can. In fact, as some animal handlers have learnt by bitter experiences, such sudden changes to routines could end up in catastrophe as the animals on a routine, could simply turn wild at the sudden break in the routine on which their minds had been fixated the whole time the game was going on.

This is because, by virtue of their three intelligences, animals only have three options when any situation suddenly presents, which are (i) the II option of having an automatic solution generated which can be instantly applied to such situation (ii) the PI, if it was one calling for routines (reason why they could be such a sport when in these sort of routines) and (iii) the CI option or basic instinct of fight or flight. Therefore, when the routine is broken, option ii, Procedural Intelligence is out, in which case the animal is left only with options I and iii. And since a broken routine does not present a novel situation, the only prompt that could enable option i, the II to generate and provide an instant solution to be applied to the situation, then option I, the II is also out; and all of a sudden, they are left momentarily clueless, but for option iiI, the CI.

The animal, with no capability of reasoning out the situation, is then left with the fight or flight option only. In which case, a number of factors will decide its next move, such as, for example, the nature of the bond between it and the handlers or how friendly it perceives the atmosphere of the venue and of course its own mood at the time, for example, as regards whether it was already getting hungry and therefore angry or not. One such case was in December 9, 2009 in Hamburg, Germany when **Christian** Walliser, 28, an experienced tiger trainer, was attacked after he stumbled during the show in Hamburg. The 200 guests watched in horror as Walliser was pinned to the ground by the tigers. The tigers dug their teeth into Walliser's head and upper body, tearing off most of his left hand. Doctors amputated Walliser's left hand and said he had suffered serious head and chest injuries in the attack. (Source—'A listing of Big Cat attacks' Yahoo news online).

Such aggressive actions by animals which we immediately categorize as animal unpredictability, with terms such as 'these are wild animals' is anything but that and such notions should now, in the light of this theory, be re-appraised.

This theory affirms, corroborated by its oh parameter which rooted in our phylogeny is self verifying, that no action by an earthling can or should be classified as unpredictable unless the organism is insane. As has already being made clear in earlier chapters, earthlings are one big family sharing a common phylogeny and imbued with concom, which means they are purposeful. Therefore a purposeful organism can hardly be classified as unpredictable or purposeless, since it has only one mission, which is to fulfill the purposes of concom, which in turn centers or is contiguous on survival. All its actions, even a wild animal, are geared towards that and are thus predictable and purposeful, once the dynamics at work are understood. Therefore, to say an organism's action is unpredictable is to say it is insane and as such unless an organism be adjudged insane, its action cannot be justifiably described as unpredictable. Understanding animals as regards how and when they engage their three different intelligences is paramount in any theory regarding their predictability.

We classify it as unpredictability only because of our poor understanding of the animal's frame of mind at the time of such attacks; a frame of mind that is only the product of its II intelligence, whose mechanism entails only instant computation and action, with no element of or a time for rationalization. Furthermore, such instant computation could be influenced by any one or the resultant of all the internal and external stimuli effective upon the organism at the time. Thus, a sudden gnawing pain in its stomach, which is not perceptible to the humans may be responsible for an animal suddenly lashing out at anyone near, or something just as imperceptible but which the better developed sub-sensory perception of the animal has just detected in the surrounding or even on or about the handler. Furthermore, animals, just like humans, have different temperaments and while docile ones may be trusted to hold their peace under certain circumstances, others will not Even humans naturally, if unwittingly, tend to lapse back, from time to time, into the CI, in rather ordinary situations, such as being on edge when stressed or being prone to anger when hungry. Therefore animals are just as predictable as human counterparts, once it is understood that they rely solely on II, whose mechanism and consequently, manner of execution is totally distinct from our RI. Intelligent decisions which humans sometimes make, involving no prior reasoning and which we refer to as intuition, are in fact nothing more than vestiges of our lost II, at work.

In comparison therefore, while our rationalization ability may be superior, it is pertinent to bear in mind that that the primordial II is no less efficient, and serve tremendously well in matters of natural survival. As a matter of fact, human failure to recognize the remarkable capabilities of II has often led to grievous mistakes being made with devastating consequences, especially by animal handlers or keepers.

A typical example and which serves as a good comparison between II and RI intelligence is the unfortunate incident between a tiger and its keeper in a UK zoo. The experienced zookeeper had wanted to retrieve an item from inside the enclosure of the tiger in question, and probably couldn't be bothered to go through the rigmarole of the standard procedure of opening and closing gate after gate in order to properly isolate the cat elsewhere to enable a safe entrance to enable the item's retrieval from the tiger's enclosure. He was safely outside the enclosure and the said item was just within reach from where he was. All he needed to do was reach out with an arm through the barrier to make a quick grab of the item and pulled his hand out before the cat could even make a move from where it laid; so his rationalizing thought process, RI, must have gone.

He, underestimating II, was confident that there was no way the animal could know or even guess that he planned to stick his hand into its cage and as such it couldn't think of reacting until he has made the dash—grab for the item and pulled his hand out safely. This is the kind of false impressions held by people who believe that animals have no minds of their own or which they use as actively as ours. Such notion was about to prove dreadful for the otherwise experienced man. This is because unknown to him, the animal too knew what may happen next, through its own mind, having the information in its memory that there was some connection between him and the item in its enclosure, probably from having seen them together in the past.

The stage was set for a rare but classical encounter between an animal and a human intelligence. While the animal did not have the luxury of reasoning like him, the difference in mechanism between its II and his RI is amply demonstrated in the way the animal's intelligence worked, much like an electronic calculator, as below illustrated:

Item in my enclosure seen together in the past with him + Him now coming close to, interested in item = item together with him inside my cage? = get stealthy for a pounce (solution)

However, the only thing the animal will ever know about the computation above is the final answer (solution). The computation had been done by a mechanism called intuition, involving no thought process. It is pertinent to understand that the cat had no capability of determining how he will be together with the item in its cage, nor can it rationalize that the only way he could achieve that, with a barrier in between them, was only by sticking in a hand. As far as its mind went, they could be together in its cage and since he was a prey item, all it was going to do was to wait and get ready to go for him. He must have felt it was fairly safe to attempt the dash—grab, after having taken into account all associated parameters, including especially the distance of the animal from him and its predisposition (it is not poised for an attack since as far as his rationalization goes, the tiger couldn't even guess what he was about to do), and therefore even if it did try to attack his hand, it would start too late, by which time his hand must have been in and safely out with the item. The cat on the other hand, watching him intently (to the keeper that may not have seem anything unusual, cats always stare intently anyway) was already poised but knew that it could not afford to alert him of that fact, as otherwise he could decide against the attempt. As such it kept still, even though every nerve of his entire body as well as its biochemistry must have strained, ready to pounce in an instant, with an advantage in sensory and sub-sensory perception, an attribute of its II. The animal at this stage, unknown to him was at its stealthiest, even though pretending to keep still. After a few more minutes of procrastinating, he rationalized it was safe enough an adventure and made the grab—dash. While he had been fooled by the tiger, who didn't betray its intention, the cat could use its II to read him, almost as clearly as you could read a clock on the wall; such is the efficiency of that totally natural, uncorrupted by rationalizations, intelligence. The moment he made up his mind, the tiger saw the hand movement, by sub sensory perception, which meant just about the exact second as he had started the reflex action movement of the hand, and was up in a pounce at the same instance, getting the hand before it could be pulled out. While this might have seem highly improbable to the rationalizing zookeeper, who was no fool before deciding to make the dash—grab and who must have been surprised at the unexpected move, his underestimation of II had proved costly. In the instantaneous calculations, whose results are nature only facilitated and as such extremely reliable, the cat has computed all information stored in its memory about the zookeeper and its own attack tactics and at the same instant made comparisons with similar or near similar past experiences, weighed up all the options available to enable it make an split—hair accurate pounce. Such assessments will include far many more parameters than our reasoning mind could attend to and come up with a rational solution in such a split second timing and since all of these will work in consonance with the animal's biochemistry, perhaps only a specially designed computer could have beaten the animal in the contest. Other parameters that would have been taken into consideration to arrive at the instant solution will include considerations of the cat's agility, the distance to be covered in the snap second pounce and exactly how much force is needed in that pounce to get the target accurately (mind you, too long and the cat will clumsily collide with the barrier and too short will miss the target outright). Furthermore, the sub-sensory perceptions will enable to be added to that data, parameters such as the slightest give away body movements that will show the hand movement had begun, such as sounds that might have been inaudible or discarded by a reasoning intelligence like those betraying a movement even before the movement becomes perceptible, e.g. the first rustle of the shirt of the zookeeper as the hand started its stretch for the grab. Even the slight increase in body odor from his armpit as he raised the arm for the stretch into the enclosure, and other such details are all parameters that would have been taken into account. All of these data will be computed by the tiger's awesome II to arrive at an instantaneous best and final solution, such that the pounce was deadly accurate; an accuracy that could never be matched by RI's comparatively sluggish and time consuming mechanism. Therefore, the tiger moved at the exact moment, with the exact velocity required, in the exact angle necessary and with the exact stretch and force in the pounce to get the hand before it could be pulled out of the enclosure, in what was supposed to be an instant dash—grab by an intelligent human. These sorts of feats are common place in II. Any one who has ever tried to kill a cockroach or a rat will discover that their movements are such that could hardly be generated by a RI, as the creature, despite the huge pile of odds against it, feigns and dodges around skillfully, while the human hits and misses repeatedly. The difference between PI and II, on the other hand, is amply demonstrated, for example, in the extensively documented, (U—Tube), duels between the cobra (PI) and the mongoose (II), in which the cobra's moves are always a routine and therefore predictable, while the mongoose's next move is anything but, as it dodges, feigns and taunts the reptile, just so to get it to bring its head down from its usually menacing upright stance, and then, in a lightning flash, lashes out so fast, with a move so unexpectedly out of the blues when it happens that it is almost imperceptible,; as it kills the snake, with a strike aimed intelligently only always specifically at the head of the cobra.

Talent

Talent, a human attribute, depicts a natural excellence, which is not reason reliant, in the art of devising a particular means and so is a good example of Intuitive Intelligence. Thus, animals, indeed other organisms may be described as totally talented in what they do, since it comes naturaly and there is no intrusion of RI. This again brings to the fore, the masterminding game between organisms, that started at the origin of living cells. The accomplishment of retroviruses, for example, will suggest that while humans may be the highest intelligence, with four different levels to boast of, organisms such as these have proved worthy adversaries, at the cellular level, by devising means to outwit humans. A better understanding of the workings of the minds and a new respect for the intelligence of fellow earthlings may be another way forward in establishing absolute human superiority in the war of survival. So called superbugs are also organisms superior in intelligence to us at the cellular level, and which like the retroviruses, we have been unable to subdue with the RI.

Other positions of this theory may be briefly summarized as follow:

The Sleep Anti-Thesis (SAT) Hypothesis of consciousness.
All living things share, each at its own level, a common consciousness which is essentially a sleep anti-thesis of consciousness; the reason why the state of coma exists, which is a state of deepest sleep, followed by 'sleep-abysmal', which is death.

The Hypothesis of Surrection.
Life surrected from death, which is zero consciousness, and evolved in a gradually increasing continuum through

different stages, broadly the plant, reptilian, mammalian to the highest, which is human consciousness.

The Consciousness-Continuum (CC) Hypothesis of Consciousness.

Consciousness exists in a gradual continuum in all life forms, the Consciousness-Continuum(CC), from the simplest to the most complex, each type occupying different slots of that continuum, with the occupants of any level possessing the attributes of lower levels.

The Delanin Fadahunsy (DF) Equilibrium of Consciousness DF EQcons.

Life is consciousness in a reversible equilibrium, the direction of which is influenced by anyone or the resultant of all the internal and external stimuli effective on an organism; with phylogeny forwards in the onward aspect, and backwards in the reverse.

The Delanin Fadahunsy(DF) definition of consciousness.

Consciousness is herein defined as response (or the possibility thereof) to stimuli, any, by a living organism.

The constituents of Consciousness.

Consciousness is made up of two elements, attention (alertness) and concentration, with the former being but a unit of the latter; while the level (intensity) of concentration in turn determines the level of consciousness.

The Age Hypothesis of Consciousness.

Barring interference, the most ancient, in phylogeny, leads the most recent, in all states of consciousness.

The Regurgitation hypothesis of Consciousness.

In consciousness, dissolution regurgitates phylogeny.

The Three Tiers Hypothesis of Consciousness.

Consciousness consists of three main classes determined by constituent nature of attention and or concentration.

3rd Class Lower (NAM) consciousness

Made of permanent attention (e.g. Algae, Plants).

3rd Class Upper (NAM) consciousness

Made up of higher intensity permanent attention bordering on low-level concentration, herein referred to as 'Non-Discernible Concentration' (e.g. Reptiles).

2nd Class (RAM) consciousness

Made up of a permanent attention underlay, followed by a medium intensity sporadic discernible concentration(e.g. Mammals).

1st Class (PAM) consciousness

Made up of a permanent attention underlay, followed by sporadic discernible consciousness, on top of which is a high intensity permanent discernible concentration (e.g. Humans).

DF Definition of self-consciousness.

Self consciousness is the act of permanently looking into one's mind (memory), which resulted from the attainment of high intensity or Permanent Discernible Concentration by the Early Hominin, in addition to the already possessed mammalian Sporadic Discernible Concentration to create a 'picture-within-a-picture' television-like capability.

The Rising Dough Hypothesis of the growth of the human brain.

Earlier in the evolution timeline, just before the brain's development, especially from the onset of the growth of the reptilian brain, the brain is comparable to a freshly kneaded wheat dough, younger and highly impressionable and waiting for the right condition, prompting stimuli, to grow, and once it has finished growing, like a fully risen dough, is most probably unable to grow any more.

Origin of self–consciousness

Self–consciousness, a phenomenon that is only gradually acquirable, originated around 2.588Mya, specifically within a bipedal line of extant apes, that turned out to be pre - humans, as a result of the extension of the span of Discernible Concentration beyond the typical mammalian sporadic; until extended to Permanent Discernible Concentration, PDC, thus enabling them to become fully self conscious not later than approximately between 500,000 and 125,000ya.

The COF hypothesis for the acquisition of definitive full self-consciousness

This theory posits, using the control of fire as a milestone, that it could be safely assumed that humans had, by the period 500,000 to 125,000 years ago acquired a permanent presence of mind or self–consciousness.

The hypothesis of intelligence as a measure of presence of mind.

Intelligence is more a measure of presence of mind, which is access to memory and which is determined by the discernible concentration span of an organism, as well as the intensity of that concentration rather than brain size.

The Age Hypothesis of Consciousness.

From ontogeny to adulthood, *Homo sapiens*, the highest consciousness, recapitulated the phylogeny of consciousness.

The Oddity Hypothesis.

Any action of an organism that is neither observable in, nor deductible, inferable or traceable from phylogeny can only be considered unnatural or insane.

The Consciousness Compulsion (Concom) Hypothesis.

Quite apart from the biochemistry of the evolution of life, exists the Survival or Consciousness Compulsion (Concom), without which, perfect biochemical possibilities regardless, conscious life was impossible as it is the inertial breaking force essential for the evolution of such life; and which later became, in its step-down form, the 'will for survival', which is just as equally crucial for its continued existence, and without which it will easier revert to the natural state of sleep 'Concom is conscious life's self propulsion mechanism.

The hypothesis of presence of mind (POM) over brain size as a measure of intelligence

Intelligence is more a measure of presence of mind, which is access to memory, which in turn is determined by the level and span of discernible concentration attainable by an organism, than it is of the brain size or brain to body weight ratio of that organism.

The Permanent Access Memory (PAM) Hypothesis of the development of the human brain

Postural habitual biped fervor and its ensuing predation consequences resulted in the gradual extension of the sporadic mammalian concentration, in human ancestral line of apes, until it eventually became permanent concentration, enabling permanent access to memory or permanent presence of mind, in the process of which the brain developed.

The impracticality of attempts to design a conscious robot.

Consequent upon the Consciousness Compulsion hypothesis, it is unfeasible to design conscious life, due to the impracticality of imbuing it with the Consciousness Compulsion and as such any designed life will always remain programmed, one way or another. Consciousness is a bio-chemical concept that was in the making for over four billion years, with everyday and every minute of those years essential in the 'assembly' of the final products of today, the earthlings; making conscious life inconceivable as an electro-mechanical or other such concept.

DF Definition of sleepwalking (somnambulism)

Sleepwalking is here defined as an non-insane automatism exercise act whose function is to raise a sleepwalker's consciousness level, so s/he does not fall through the consciousness—continuum to unconsciousness or coma.

'Controlled Artificially Induced Rem Atonia Techniques' (CAIRAT) as treatment for DFD and DFD manifestations

Possibility of using some form of 'Controlled Artificially Induced Rem Atonia Techniques' (CAIRAT) with caution in the treatment of DFD (Sleepwalking) and DFD manifestations (RBD), at those rare instances when intervention may become imperative.

That children suffer proportionally more from DFD than adults is only natural

More children suffer from DFD than adults because they are still at a stage when they are just developing their PDC consciousness and as such rely more on the two other consciousnesses, the reptilian (NDC) and the mammalian (SDC), with the result that they more easily, at sleep, lapse back to the deep end of the reptilian consciousness at which DFD occurs.

Self-Cognition in animals.

Self-cognition in animals does not equate to self-awareness or self-consciousness.

Mind as memory.

The mind is hereby defined by this theory as an organism's entire database of information, including all about self and its world, stored bio–chemically, as memory. Therefore, animals like humans have minds of their own, but while humans are more predisposed to engaging theirs through a thought mechanism called rationalization, animals, having no such level of intelligence, engage theirs through a mechanism called intuition.

Concerning perception.

The degree of reward of our perceptions is directly related to the difference in degree between our PDC (the larger screen of our 'picture-within-a-picture consciousness, our entire memory database), and how (in what way distinguished) and/or what our SDC (the in-picture) perceives; the higher the difference in degree, the more emotionally rewarding or otherwise the experience.

The 'Arboreal Gymnastics Escape' (age) hypothesis.

'Age', introduced by this theory, should take primacy over speed, in the pros and cons of bipedalism and other such considerations in primates, especially in wooded habitats.

Sleepwalking as an non-insane automatism.

In the naturally occurring unconscious state of non-automatism, called sleepwalking, the use of muscles is, directed, pre-programmed by nature specifically, at performing the function that nature assigned to sleepwalking, and could therefore be no longer redirected for any other simultaneous use, anymore than a nose could be redirected to smell a flower, at the exact instance of a sneeze simultaneously.

DF Definition of Dreams.

Dreams is Partial Virtual Memory Recall (PVMR), which occurs when the sleeper is at a high enough consciousness level to access memory, usually closer to wake up level (REM) than to deep sleep (NREM) consciousness.

Adaptation and Intelligence

Adaptation is an intelligent process in which an organism devises means to adjust to its environment, involving, coordination, through communication, of all intelligences available to such organisms, resulting where essential in adaptive traits. Intelligence as already defined by this theory means the art of devising means and therefore without intelligence, there can be no adaptation. In lower organisms, adaptation is achieved using physiological or other such sensors to interact with the environment and data so collected are communicated, indicating areas of need in a prioritized manner, to the organism's Cellular Intelligence, which then devises the means to meet such needs, with traits appearing at the areas of greatest needs, in accordance with the analysis of such data and using such as specification for the design of such traits; in much the same fashion as an architect uses data collected to fashion a home design to meet specific needs. In higher organisms such as mammals, Intuitive Intelligence interacts with the environment, through which such data are collected and then communicated to the organism's Cellular Intelligence serving, in similar fashion as in lower organisms, to guide the development of adaptive traits based on prioritization of needs, directing the division of cells to meet those specifications. There is no reason to suggest that the latter addition Reasoning Intelligence plays any direct part in natural adaptation.

Therefore adaptation is not just a result of organisms' response to environmental signals, but rather the peculiar response of each organism to such signals, as dictated by the organism's intelligence. Thus, the giraffe's art of devising the means for food involves peculiarly repeatedly stretching its neck for food on tall trees. This is the way its intelligence has, in its own way resolved the problem of finding food, and such is then communicated over time to its Cellular Intelligence, like data collated, acting as specifications and guiding the division and growth of cells to meet the specific way it has chosen to devise the means of finding food, giving rise to its long neck. However, a similar organism such as the zebra, for example, may choose to devise a different means to achieve the same purpose, even thought they may co–habit the same environment, and such a different means is also communicated to its Cellular Intelligence, guiding the development of traits to meet its own peculiar way of trying to solve the same problem, thereby giving rise to a different trait, in the solution of the same problem; whereas if intelligence was not involved, then organisms would be expected to react to environmental signals similarly, in a robotic kind of manner, and traits will be more or less the same, i.e. both the giraffe and the zebra would be expected to have long necks. Thus, each individual organism will use its own level of intelligence to devise its own means, culminating in variants and species. However earthlings, being concompelled, as explained earlier, are imbued with making progress only, and therefore organisms can only keep getting smarter. Fitness is therefore not unlike a measure of intelligence and therefore organisms with the worse ideas or contributions, in the art of devising means, will eventually lose out in the game of survival. In other words, natural selection ends up eventually culling the less intelligent organisms, such that over time there is the preponderance of the smarter ones and such play of events basically extrapolates. Life then may not actually be the survival of the fittest, but rather the survival of the smartest, as only organisms that are more intelligent (better in the art of devising means and therefore fitting more) rather than just the fittest survive better. It's the only plausible explanation for how, for example, our ancestral line of apes were able to survive as bipeds while co–habiting the same environment with the big cats, as earlier explained in the theory. It is also why both preys and predators are getting smarter e.g. foxes and rabbits running faster than their parents. Perhaps no place is this better demonstrated than in the home front where traps, once very reliable, are now almost entirely useless, as rats now avoid them cleanly all the time, regardless of how tempting the bait; and in corroboration of the fact that the level of intelligence will always keep rising, they now can also manage to distinguish between safe and any poisoned foods, meant to kill them.

The scare stunt hypothesis of the origin of bipedalism.

Bipedalism originated as a scare stunt, just like the upright stance of the cobra, most probably instinctively and got retained initially only because of its effectiveness as such, before its different later adaptations.

DF Definition of Intuition.

An innate mechanism, in a cognitive organism, prompting automatic computation of the best instantaneous solution, whose quality is time independent, to a problem, using the organism's mind (memory database) and perception, much like an electronic calculator computes instant solutions, without any thought process involved

The Four Tiers Intelligence (FTI) hypothesis.

This theory postulates that living organisms occupy one of four levels of intelligence, which in the order of increasing complexity are Cellular, Procedural, Intuition, and Reasoning, with organisms in a higher level possessing, in varying degrees, the attributes of lower levels. Thus, humans for example have four intelligences, other mammals three, etc.

The postulate of reason over intelligence.

This theory postulates that it is more the ability to reason rather than the level of intelligence that sets the human race apart from animals, as it is rational to suggest that animals could live a settled and civilized life, even with their level of intelligence, if only they had the ability to reason; especially since they appear to be cranially capable of even more complex tasks than required for such undertaking.

DF Generic definition of intelligence.
Intelligence is the art of devising means

DF Specific Definition of intelligence.

Intelligence is here defined as any innate physiological or other such natural mechanism or process, involving the use of the mind, imbuing the capacity to initiate solutions to problems, internal or external, in a living organism. Thus, for example, our natural defense mechanism, called the immune system, which initiates solutions to internal threats, is a form of intelligence, the primordial Cellular Intelligence (CI).

DF Generic definition of mind.

The mind is defined as the database of the entire information potential, intrinsically, biochemically, stored in an organism. In higher intelligence, part of such storage is centralized in the brain. Therefore, if intelligence is the capacity to initiate solutions, then the mind is the obligatory tool, without which there can be no initiation.

Talent

Talent in humans is a natural aptitude, not attributable to Reasoning Intelligence, but to Intuitive Intelligence, depicting an excellence in the art of devising a particular means, which may then be expressed with or without the assistance of the Reasoning Intelligence. Talent is thus another example of II in humans.

Mind in all.

All living things are intelligent and as such all have a mind of their own.

Actions of animals are as predictable as that of humans.

This theory postulates that while humans employ the Reasoning Intelligence, animals employ Intuitive Intelligence, therefore no action of an animal can be adjudged unpredictable, unless such animal be adjudged insane. An animal's behavior according to this theory, at any given moment, can be influenced by anyone or the resultant of the external and internal stimuli effective on the animal, at that particular moment; the understanding only of which in addition to that of the mechanisms and engagement of its three Intelligences, will guarantee a good measure of the predictability of its behavior.

Postural habitual biped fervor (pohabifer) hypothesis of the adoption of postural habitual bipedalism in human ancestral line of extant apes.

This theory postulates that while bipedalism almost certainly originated instinctively as a scare stunt with later adaptation such as occasional bipedalism, being due probably to sexual or other such selection pressures, adoption as habitual posture by human ancestral line of extant apes was due to *pohabifer*, a peculiar stubborn streak, which made them specially predisposed to the motivational senses of their mammalian brain.

Copyright @ Delanin Fadahunsy 2011

Bibliography

1. A, Haviland, HalandE.N.Prins, Dana Walrath and Barry McBrd in 'Evolution &Podostomy;the human challenge'(1998)
2. Aerts, Peter; Evie E. Vereeckea, KristiaanD'Aoûta (2006). "Locomotor versatility in the white-handed gibbon (Hylobateslar): A spatiotemporal analysis of the bipedal, tripedal, and quadrupedal gaits". *Journal of Human Evolution* (Elsevier) **50** (5): 552-567. doi:10.1016/j.jhevol.2005.12.011. PMID 16516949.
3. Barlow, David H. and V. Mark Durand, *Abnormal Psychology: An Integrative Approach*, Cengage Learning, 2008, ISBN 0495095567, p. 300.
4. Bingham, Roger; Terrence Sejnowski, Jerry Siegel, Mark Eric Dyken, Charles Czeisler, Paul Shaw, Ralph Greenspan, Satchin Panda, Philip Low, Robert Stickgold, Sara Mednick, Allan Pack, Luis de Lecea, David Dinges, Dan Kripke, GiulioTononi (February 2007). "Waking Up To Sleep" (Several conference videos). The Science Network. Retrieved 2008-01-25.
5. ^Bauer, Harold (1976). "Chimpanzee bipedal locomotion in the Gombe National Park, East Africa". *Primates***18**: 913. doi:10.1007/BF02382940

6. CNN,Sleepwalking defense in Arizona murder trial,May 25, 1999"

7. Brunet, Michel; Guy F, Pilbeam D, Mackaye HT, Likius A et al. (11). "A new hominid from the Upper Miocene of Chad, Central Africa". *Nature* **418**: 145-151. doi:10.1038/nature00879. PMID 12110880.

8. Burk, Angela; Michael Westerman and Mark Springer (September 1988). "The Phylogenetic Position of the Musky Rat-Kangaroo and the Evolution of Bipedal Hopping in Kangaroos (Macropodidae: Diprotodontia)". *Systematic Biology* **47** (3): 457-474. doi:10.1080/106351598260824. PMID 12066687.

9. Butler, A. B. and Hodos, W. Comparative Vertebrate Neuroanatomy: Evolution and Adaptation, Wiley

10. deBenedictis, Tina, PhD; Heather Larson, Gina Kemp, MA, Suzanne Barston, Robert Segal, MA (2007). "Understanding Sleep: Sleep Needs, Cycles, and Stages". Helpguide.org. Retrieved 2008-01-25.

11. deBruxelles, Simon (18 November 2009). "Sleepwalker Brian Thomas admits killing wife while fighting intruders in nightmare". *The Times* (London).

12. ^Dhingra, Philip (2004-05-25). "Comparative bipedalism: How the rest of the animal kingdom walks on two legs". Retrieved 2007-10-29.

13. Djawdan, M (1993). "Locomotor performance of bipedal and quadrupedal heteromyid rodents". *Functional Ecology* (British Ecological Society) **7** (2): 195-202. doi:10.2307/2389887.

14. Djawdan, M.; Garland, T., Jr. (1988). "Maximal running speeds of bipedal and quadrupedal rodents". *Journal of Mammalogy* (American Society of Mammalogists) **69** (4): 765-772. doi:10.2307/1381631. JSTOR 1381631.

15. Dreifuss, Fritz, and J. Penry. "Automatisms Associated With the Absence of Petit Mal Epilepsy." *Archives of Neurology* 21.2 (1969):142-149.

16. ^Fleagle, J.G (1994). "Primate locomotion and posture". In Steve Jones, Robert Martin & David Pilbeam. *The Cambridge Encyclopedia of Human Evolution.* Cambridge: Cambridge University Press. pp. 75-79. ISBN 0-521-3270-3. Also ISBN-0-521-46786-1 (paperback)

17. ^Garland, T., Jr. (1983). "The relation between maximal running speed and body mass in terrestrial mammals". *Journal of Zoology, London***199**: 157-170. doi:10.1111/j.1469-7998.1983.tb02087.x.

18. Handwerk, Brian (2006-01-26). "Dino-Era Fossil Reveals Two-Footed Croc Relative". National Geographic. Retrieved 2007-10-29.

19. Haviland, Haland E.N.Prins, Dana Walrath and Barry McBrd (1991) 'Evolution &Podostomy; the human challenge'.

20. Hayward, T. (1997). The First Dinosaurs. *Dinosaur Cards.* Orbis Publishing Ltd. D36040612.

• ^Hodos, William. "Comparative Vertebrate Neuroanatomy: Evolution and Adaptation Wikipedia the free on line encyclopedia

- Huffard CL, Boneka F, Full RJ (2005). "Underwater bipedal locomotion by octopuses in disguise". *Science***307** (5717): 1927. doi:10.1126/science.1109616. PMID 15790846.

- Humphrey, N., Skoyles, J.R., and Keynes, R. (2005). "Human Hand-Walkers: Five Siblings Who Never Stood Up" (PDF). Centre for Philosophy of Natural and Social Science, London School of Economics.

- Hutchinson, J.R. (2006). "The evolution of locomotion in archosaurs". *ComptesRendusPalevol***5** (3-4): 519-530. doi:10.1016/j.crpv.2005.09.002.

- Kales, A. and others. 1980. Hereditary factors in sleepwalking and night terrors. The British Journal of Psychiatry 137: 111-118.

21. Kazlev, et al., M. Alan (2003-10-19). "The Triune Brain.". *KHEPER*. Retrieved 2007-05-25.

22. Keith Oatley, DacherKeltner, Jennifer M. Jenkins. Understanding Emotion (2006) Second Edition. Page 235.

23. Kudrimoti, H.S.; Barnes, C.A.; McNaughton, B.L. (1999). Reactivation of hippocampal cell assemblies: Effects of behavioral state, experience, and EEG dynamics [Electronic version]. *The Journal of Neuroscience, 19*, 4090-4101.

24. Lederman, Eliezer. "Non-Insane and Insane Automatism: Reducing the Significance of a Problematic Distinction." *The International and Comparative Law Quarterly* 34.4 (1985): 819.

25. Lyon, Lindsay. "7 Criminal Cases that Involved the 'Sleepwalking Defense.'", *US News and World Report*. May 2009.

26. NCHeglund, GA Cavagna and CR Taylor 1982 Energetics and mechanics of terrestrial locomotion. III. Energy changes of the centre of mass as a function of speed and body size in birds and mammals Journal of Experimental Biology 97:1

27. McHenry, H.M (2009). "Human Evolution". In Michael Ruse & Joseph Travis. *Evolution: The First Four Billion Years*. Cambridge, Massachusetts: The Belknap Press of Harvard University Press. p. 263. ISBN 978-0-674-03175-3.

28. Mendelson, Wallace, Christian Gillin, and Richard Wyatt. "The Sleep Stages." *Human Sleep and its Disorders*. New York: Plenum, 1977. 3-10.

29. ^Panksepp, J. (2003). Foreword to Cory, G. and Gardner, R. (2002) *The Evolutionary Ethology of Paul MacLean*. Greenwood Publishing Group.

30. ^Patton, Paul (December, 2008). "One World, Many Minds: Intelligence in the Animal Kingdom". *Scientific American*. Retrieved 29 December 2008."

31. Psychology World (1998). "Stages of Sleep" (PDF). Retrieved 2008-06-15. "(includes illustrations of "sleep spindles" and "K-complexes")"

32. Rose, M.D. (1976). "Bipedal behavior of olive baboons (Papioanubis) and its relevance to an understanding of the evolution of human bipedalism". *American Journal*

*of Physical Anthropology***44** (2): 247-261. doi:10.1002/ajpa.1330440207. PMID 816205.

33. Sharma, Jayanth (2007-03-08). "The Story behind the Picture—Monitor Lizards Combat" (php). Wildlife Times. Retrieved 2007-10-29.

34. ^Schmitt, Daniel (2003). "Insights into the evolution of human bipedalism from experimental studies of humans and other primates". *Journal of Experimental Biology* **206** (Pt 9): 1437. doi:10.1242/jeb.00279. PMID 12654883.

35. ^Sharp, N.C.C. (1997). "Timed running speed of a cheetah (*Acinonyxjubatus*)". *Of Zoology, London***241**: 493-494. doi:10.1111/j.1469-7998.1997.tb04840.x.

36. Rachel Nowak (2004-10-15). "Sleepwalking woman had sex with strangers". New Scientist. Retrieved 2007-04-30.

37. Rechtschaffen A, Kales A, editors. A Manual of Standardized Terminology, Techniques and Scoring System for Sleep Stages of Human Subjects. Washington: Public Health Service, US Government Printing Office; 1968.

38. Suwa, Gen; Kono RT, Simpson SW, Asfaw B, Lovejoy CO, White TD (2). "Paleobiological implications of the Ardipithecusramidus dentition.". *Science***326**: 94-99. doi:10.1126/science.1175824. PMID 19810195.

39. Swanson, Jenifer, ed. "Sleepwalking." *Sleep Disorders Sourcebook*. MI: Omnigraphics, 1999. 249-254, 351-352.

40. T. Douglas Price, Gary M. Feinman (2003). *Images of the Past, 5th edition*. Boston: McGraw Hill. pp. 68. ISBN 978-0-07-340520-9.

41. University of Liverpool-Research Intelligence Issue 22—Walking tall after all

42. Waldman, Dan (2004-07-21). "Monkey apes humans by walking on two legs". MSNBC. Retrieved 2007-10-29.

43. Walker, M.P. "The Role of Sleep in Cognition and Emotion." Annals of the New York Academy of Sciences. 1156. (2009): 174.

44. Winifred Henke and ThorroffHardt Handbook of Paleoanthropology (1997)

GLOSSARY

AGE/ATE	- Arboreal Gymnastics Escape/Arboreal Type Escape
CC	- Consciousness–Continuum
Concom	- Consciousness Compulsion
DF	- Delanin Fadahunsy
EQCons	- Equilibrium of Consciousness
Mens rea	-'mens rea is Latin for 'guilty mind'. A person cannot be held responsible for a crime if his/her mind is not guilty of the crime. In other words, an individual cannot be held responsible for a crime, if it was discovered that they were insane at the time the crime was committed. This is because even though they committed the crime, they were unaware of it, and as such had no 'guilty mind' or 'mens rea'. The standard common law test of criminal liability is usually expressed in the Latin phrase 'actus nonfacit reum nisi mensa

sit rea', which means 'the act does not make a person guilt unless the mind be also guilty'

NAM	- Non - Access Memory
NDC	- Non Discernible Concentration
PA	- Permanent Attention
PAM	- Permanent Access Memory
PDC	- Permanent Discernible Concentration
SDC	- Sporadic Discernible Concentration
Sleep–Abysmal	- Zero consciousness/Death
Surrected	- Came to life from sleep-abysmal

Lightning Source UK Ltd.
Milton Keynes UK
UKOW021017291111

182885UK00001B/2/P